'A rallying cry against the rise ... e era we are moving into, we are in danger of bec ... algorithms and the rise of anti-humanism. We must res ... Martin, *The Times*

'The best analysis of the contemporary situation I have ever read' Zoe Williams

'A very interesting book, wide-ranging, insightful and yet still optimistic . . . he is passionate in his defence of enlightened humanity against the current two-pronged onslaught from the politics of fear, hate and unreason, and the tech-driven dreams of the super rich' Ed O'Loughin, *Irish Times*

'Passionate and persuasive . . . deep, critical and lively . . . illuminating, impressive' Mark O'Connell, *New Statesman*

'Amid the ruins of many modern ideologies, Paul Mason's consistently bracing book offers a guide to a sustainable future – one that we can still shape with a fresh transformational vision of what it means to be free human being. Everyone should read it' Pankaj Mishra

'An unshakable humanist faith runs through this book . . . with his humane stress on the good life, Mason defies the caricature of the Corbyn left as reheated Soviet Communism. Corbynism is also routinely charged with wanting to "take us back to the 1970s". But here its leading thinker engages with tomorrow's economy with an urgency that's not currently matched on the right' Tom Clark, *Prospect*

'An undeniably compelling account of the apocalyptic dangers that await us as we hurtle into the 21st century' Jamie Maxwell, *The Herald*

'Unlike other recent works with an accelerationist bent, Mason seeks to bolster his techno-utopian proposals with a robust ethical framework. Admirably, Mason's books never shy away from tackling big ideas and making them accessible to a general audience . . . a bracing intervention in the current culture wars . . . imbued with a refreshingly pragmatic optimism too often absent from leftist analyses' Calum Barnes, *Morning Star*

'Persuasively argued, clearly considered, compelling stuff . . . it is a fantastic book . . . you may not agree with him on everything but he is doing the thinking needed to address some of the most ...

'In ...

ABOUT THE AUTHOR

Paul Mason is an award-winning writer, broadcaster, and film-maker. Previously economics editor of Channel 4 News, his books include *PostCapitalism*, *Why It's Kicking Off Everywhere: The New Global Revolutions*; *Live Working Die Fighting*; and *Rare Earth: A Novel*.

PAUL MASON

Clear Bright Future
A Radical Defence of the Human Being

PENGUIN BOOKS

PENGUIN BOOKS

UK | USA | Canada | Ireland | Australia
India | New Zealand | South Africa

Penguin Books is part of the Penguin Random House group of companies
whose addresses can be found at global.penguinrandomhouse.com

First published by Allen Lane 2019
Published in Penguin Books 2020
001

Copyright © Paul Mason, 2019

The moral right of the author has been asserted

Typeset by Jouve (UK), Milton Keynes
Printed and bound in Great Britain by Clays Ltd, Elcograf S.p.A.

A CIP catalogue record for this book is available from the British Library

ISBN: 978–0–141–98672–2

*To the memory of my mother,
Julia Lewis (1935–2017)*

The experience of my life . . . has not only not destroyed my faith in the clear, bright future of mankind, but on the contrary has given it an indestructible temper.

Leon Trotsky[1]

Contents

CONTENTS

PART IV

Marx

PART V

Some Reflexes

Introduction

By the end of reading this book I want you to make a choice. Will you accept the machine control of human beings, or resist it? And if the answer is resist, on what basis will you defend the rights of humans against the logic of machines?

In the twenty-first century, the human race faces a new problem. Thanks to information technology, vast asymmetries of knowledge have opened up – creating vast asymmetries of power. Through the screens of our smart devices, both corporations and governments are becoming adept at exerting control over us via algorithms: they know what we're doing, what we're thinking, can predict our next moves and influence our behaviour. We, meanwhile, don't even have the right to know that any of this is going on.

And that's just the nightmare of the present. In the future, as artificial intelligence develops, it will become very easy for us to lose control of information machines altogether.

An algorithm is simply the instructions for solving a problem, devised by a human and written down. For example: when I present my passport, border control knows that, if my fingerprints match the ones stored on file, they should let me through; if they don't, I get detained for further questioning.

A computer program is an algorithm running without human intervention. In one sense it is just the latest achievement in a long process of automation. For the past 200 years one of our most successful strategies has been to move workers 'to the side' of an industrial process; to make them observers rather than controllers, giving machines temporary and limited autonomy. What we do with computers and information networks is only an extension of what we did with windmills, cotton-spinning machines and the combustion engine. But once machines can give themselves instructions, the risk is that humanity steps 'to the side' permanently, surrendering control.

Millions of people have become alert to the dangers of algorithmic control. But they assume it is a problem for an ethics committee, a tech conference, a science magazine – or for the next generation to solve. In fact, it is intimately connected to the urgent economic, political and moral crises we are living through now.

Here's why . . .

Suppose I told you there was a machine that could run the country better than the government, think more logically than any single human and run autonomously? Suppose I asked you to hand control of all the important decisions in your life to that machine? Suppose I said you would be happier if you changed your behaviour to anticipate what the machine decides? I hope you would scorn the whole idea.

But try substituting the word 'market' for the machine. For three decades, millions of people have allowed market forces to run their lives, shape their behaviour and overrule their democratic rights. There is even a religion dedicated to worshipping this machine's power and control: it's called economics.

By elevating the market to the status of an autonomous, super-human spirit guide during the past thirty years we have, potentially, prepared ourselves to accept machine control sometime during the next hundred years.

During the free-market era we learned to celebrate the subjection of human beings to market forces. We treated concepts like citizenship, morality and 'agency' (the power to act) as if they were irrelevant to the workings of the world, which was now run only by consumer choice and financial engineering.

Now, however, the free-market system has imploded. The logic of selfishness, hierarchy and consumerism no longer works. As a result, the religion of the market has given way to older gods: racism, nationalism, misogyny and the idolization of powerful thieves.

As we approach the 2020s, an alliance of ethnic nationalists, woman-haters and authoritarian political leaders are tearing the world order to shreds. What unites them is their disdain for universal human rights and their fear of freedom. They love the idea of machine control and, if we let them, they will deploy it aggressively to keep themselves rich, powerful and unaccountable.

It is not too late to stem the chaos and disorder, to stop the attempt to impose new biological hierarchies based on race, gender and nationality, and to refuse machine control. But the arguments for surrendering to them are all around us.

The idea that 'humanity is already over' is deeply embedded in modern thought, from the alt-right to the academic left. No matter how much you, personally, are trying to live by 'human values', the consensus is – from Silicon Valley to the HQ of the Chinese Communist Party – that human values have no foundation; that there is no such thing as human nature, no logical basis to privilege humans over all machines, no rationale for universal human rights.

With hindsight, free-market ideology looks like the gateway drug for a more pervasive anti-humanism. And we're about to find out just how damaging this harder drug can be.

'Compete and acquire' was the first commandment of the free-market religion. In the era of de-globalization and right-wing nationalism it will become: compete, acquire, lie, control and kill. If we don't place the new technology of intelligent machines under human control, and programme them to achieve human values, the values they will be designed around are those of Putin, Trump and Xi Jin Ping.

So I have written this book as an act of defiance. When you've read it, I hope you will begin to make acts of defiance yourself. They can range from bringing down dictators, to setting up human-centred projects in your neighbourhood, to simply defying machine logic in your daily life.

To resist effectively we need a theory of human nature that can survive in conflict with free-market economics, machine worship and the anti-humanism of the academic left.

We need, in short, a radical defence of the human being.

PART I

The Events

What the mob wanted, and what Goebbels expressed with great precision, was access to history even at the price of destruction.

Hannah Arendt[1]

I

Day Zero

Ross sprints past me, his camera rolling. He taps me on the shoulder and starts to speak but I point to the GoPro taped to my crash helmet and silently mouth the word 'live' – meaning 'don't say anything that could incriminate us'. Last time we filmed a riot together was in Istanbul. This is different.

Seconds later it is Brandon who hi-fives me as he weaves through the chaos, also filming. We've criss-crossed the riot world since 2011: Cairo, Athens, Istanbul. We extend our non-camera hands and grip fingers for a millisecond. Windows are getting shattered. An SUV is on fire. Flash-bangs thump the air and the CS gas is drifting.

About a thousand young people, masked up and dressed in black, are swarming across the city with the riot police in pursuit. And by total coincidence, in this few square metres of urban battlefield, we find each other: me, Ross and Brandon – veterans at filming countries that are going to shit.

The date is 20 January 2017. The place is Washington DC. The social war that's been raging at the edges of the global system has just arrived at its centre. We are two blocks away from the White House. Donald Trump's presidency is one minute old.

As the riot gathers momentum the police are clueless: they are trained for situations where people either obey them or get shot. Today neither shooting nor obedience is possible. So they jog breathlessly behind the protesters, their bodies weighed down by pointless equipment and bloated by a lifestyle of militarized sloth. When a girl pushing a bike trips over, accidentally taking three cops to the floor, some other cops rush up to baton her, and the bike itself, while others try to help her up. The soundtrack is classic riot music: police bullhorns; radios sizzling with panicked orders; the glass of a Starbucks window smashing; young Americans chanting 'No fascist USA!'

Eventually the cops attack, the CS gas vomiting out of their half-inch hoses. Instead of fleeing, some youths in black balaclavas form a tight wedge, black umbrellas opened horizontally for protection, and rush the police line. One protester, unmasked, lies face down on the tarmac as a cop pulls a taser on him. About twenty years old, he has blond curly hair: his face betrays not one single flicker of fear. He looks at the cop, and at the camera lenses zooming on him, and states calmly: 'Fuck Donald Trump. Fuck Donald Trump.'

As the riot breaks into fragments, the cops begin chasing small groups across the city. Everything intensifies: we sprint past the American Development Bank, past Joe's Stone Crab, past the soul-drained office blocks where Washington's lobbyists work. And as we run, this act of panicked flight from a slow, unthinking enemy – across the shattered landscape of normality – reminds me of something in the movies. But I can't place it.

The night before Trump is inaugurated I meet a 72-year-old farmer from Tennessee. 'What d'ya think'a that?' he says, jerking his head towards the words 'Fuck Trump' chalked onto a path in Franklin Square. He's wearing a thick, red cowboy shirt and a pained expression. Gazing at the demonstrators, who have gathered around a thrash metal band, he mutters: 'They don't want to work. They're sick.' Which is weird, because most of the demonstrators are clearly middle-class kids with degrees and jobs.

'Know what it costs?' he continues. 'Fifty dollars for a baseball cap. Hundred-fifty for a pair o' sneakers.' Again this remark seems strange, because – being mainly anarchists – almost none of the protesters are wearing branded baseball caps or sports shoes. 'All they want is mo-ney,' he pronounces the last word as whine, stroking his outstretched palm like a beggar. His face screws up as though he's smelled shit.

And only now do I realize: he is not actually seeing the demonstrators but – in his mind's eye – the people they remind him of: poor black people in Tennessee. 'You see 'em coming outta the supermarket . . .' his eyes stiffen and bulge with anger . . . 'white t-shirt twenty dollars, sneakers hundred-fifty . . .' He is an expert on the price of the clothes black people wear.

When I try to object, his brain flips to another topic: climate change, which he believes is fake. 'Cows fart,' he exclaims, 'but now they say I gotta pay a methane tax?' He tells me that beneath the Antarctic there is a fossilized tropical forest containing the skeletons of camels, and that this proves climate change is temporary: 'What goes around comes around.'

As Washington fills up for the Inauguration I meet individuals like this on every corner. Trump has empowered them, and the US media has granted them permission to unleash what they want to unleash most: hatred. As one self-pitying racist after another unloads their story on me, it becomes clear what we are dealing with: people who've lost their power to compute logic, but for whom all the minor injustices and inconveniences in life are linked to an imagined threat posed by blacks, gays and liberated women.

We are asked by liberal commentators to understand what motivates such individuals: the economics that has impoverished them and the social change that has disoriented them. We are asked to sympathize with the unfulfilled lives they live among the motels and flyovers of the mid-West.

I prefer a harsher form of sympathy called reason, logic and science.

Asked to understand the problems of the 'white working class' I say, with the confidence of someone born white, and raised in a tough, English coal-mining town: it does not exist. 'White working class' is an identity constructed by rich people to oppress poor people, just as the identities of the 'coolie' and the 'savage' were constructed by colonial settlers to justify treating their victims as subhuman.

Let's confront the problem. If you want peace, freedom and social justice, people like the Antarctic Camel Guy are your enemies. They put a man in power – in the most powerful nation on earth – who is a racist and a tax-dodger, and who had bragged about 'grabbing women by the pussy'. In doing so they knowingly attempted to destroy the multilateral system known as globalization, reverse fifty years of progress on minority and women's rights, and replace the rule of law with that of a kleptocratic dynasty.

And such people are on the offensive in every continent. There's the

Patriot Prayer demonstrators in Portland Oregon, calling for the heads of migrants to be 'smashed against the concrete'; there's the trolls from the ruling A K Party in Turkey, sending coordinated rape threats to female journalists; there's the mobs attacking Pride marches in Russia; and the neo-Nazis spouting Islamophobic rhetoric from the podium of the German Bundestag. In India they are among the 'cow vigilantes' lynching Muslims while Prime Minister Narendra Modi – the Hindu Trump – refuses to lift a finger. In Brazil they are the footsoldiers of Jair Bolsonaro, the fascist president elected in 2018, who said of a female opponent that she was 'not worth raping' and suggested that black *quilonbolas*, the descendants of rebel African slaves, were 'not fit to procreate'.

On a wider level, their mental garbage is polluting the thoughts and social media timelines of rational individuals all over the world.

Opinion pollsters have dubbed their mindset 'authoritarian populism'.[1] They are united in opposition to human rights, which they see as rights for somebody else; to migration, which they see as polluting 'their' culture; and to all forms of multilateralism in global politics and economics, which they see as restraining the hand of a justifiably repressive state. If that was all they believed in, we could tell ourselves this is simply a surge in the kind of reactionary sentiment that always lurks within fast-changing societies.

But there is something deeper going on: a hostility to science, logic and rationality, which have been the guiding values of market-based societies for the past 500 years. As we shall see, whether or not the activists of the alt-right fully understand it, this attack on reason was theorized in advance by a section of the elite in crisis.

The eruption of learned stupidity into global politics is all the more terrifying because it's happening in the most information-rich era in history. We need to understand this situation, and work out ways of persuading as many conservative-minded people as possible to embrace rationality, restraint and the norms of democratic behaviour.

Where they cannot be persuaded, however, we have to resist them. They have declared war on evidence-based policymaking, prudence and a global system based on rules instead of naked force. Those who want to defend these values need to fight back.

To do this, we have to arm ourselves with more than just facts. We need, as the philosopher Tzvetan Todorov wrote, surveying the struggle against totalitarianism in the twentieth century, both hope and memory. But to remember what and to hope for what?

It wasn't long ago, in the early 1990s, that perfectly rational individuals believed history had 'ended'; that liberal democracy and free-market capitalism were states of perfection, making future upheavals impossible.

Since 2008 that illusion has collapsed. The financial crisis unleashed by the bankruptcy of Lehman Brothers has spiralled into a crisis of legitimacy for the free-market system, which has now turned into an attack on democracy and human rights and is placing new strains on the geopolitical system.

Trump rules America. Brexit has triggered the breakup of the European Union. The social media are awash with anti-Semitism, Islamophobia, fantasies of white supremacy and male victimhood. In Turkey, hundreds of journalists are in jail. In the Philippines, the president revels in the work of death squads. The Syrian War, which started with teenagers scrawling graffiti against Bashar al-Assad, has killed 470,000 and displaced 10 million people.[2] Over the next decade China is gearing up to place its 1.4 billion citizens under absolute digital surveillance and control.[3] This is not some dystopian fantasy from a graphic novel. It is reality.

As a journalist I used to envy the certainties of my younger colleagues, who'd been taught in the world's elite universities that the era of systemic crisis was over. I, by contrast, had spent my early twenties in Thatcher's Britain – an era of conflict, recession and social disintegration. They, it seemed, would know only cool, calm, technocratic progress.

Now, I pity them. They are being forced to watch dramatic, unthinkable events cascade across their newsfeeds each morning, for which they have no theory. Trump flies to Moscow to side with Putin against the FBI. Austria's respectable conservative party switches overnight from an alliance with socialists to an alliance with neo-fascists. In China, Xi Jin Ping breaks with thirty years of consensus government and seizes total power. Private intelligence agencies we never knew existed turn out to be manipulating elections on behalf of the highest bidder.

Because it is happening to us in real time, and seen through devices in our pockets, this new global disorder is creating a bipolar response: hyper-sensitivity to the chaos but a mood of resignation when it comes to the possibility of ending it.

As for liberalism, once the dominant ideology of the Western world, it too has become bipolar. Among educated people it is routine to hear technological euphoria expressed alongside geopolitical despair: dark foreboding about what comes after Trump alongside business plans which assume a hi-tech, automation-driven, green future. Interrogate this attitude and the assumption is, even now, that something called the Fourth Industrial Revolution will put everything right.

The argument of this book is that it will not. To unlock the potential of new technologies to boost human wellbeing, there has to be something human left to defend. But each of the crises we face – economic, geopolitical and technological – is rooted in the erosion of what it means to be human.

Since the 1980s, free-market ideology has attacked our right to possess a self that is more than a collection of economic needs. As globalization falls apart, the very idea of rights that are universal and inalienable has come under attack. Meanwhile, technology has begun to undermine our ability to act autonomously, free of digital control and surveillance: we are increasingly subject to forms of algorithmic control that we are not allowed to see, nor to understand.

None of this is accidental: as I will show in the course of this book, overt theories of anti-humanism are today stronger than at any time in the past 200 years.

I believe, despite the fear and cruelty of the present, we can still achieve what the Russian revolutionary Leon Trotsky once called 'the clear, bright future' of humankind. But as well as demystifying the sources of economic crisis and deepening our understanding of democracy, we need to defend the very concept of humanity and draw new practical conclusions from it.

After we'd escaped the police on Trump's inauguration day, I remembered what the scene reminded me of: a zombie movie. The first zombie movie appeared in 1932, but the genre remained niche until the 1960s.[4] In most of the early zombie flicks the monster is a re-animated black Caribbean man intent on ravaging white women. It's not hard to work out what fears those films were playing on.

Only in *Night of the Living Dead* (1968) did we meet the modern zombie: a corpse brought back to life, programmed to kill human beings and eat them. This new kind of monster is just your ordinary white neighbour gone crazy. After that the zombie movie went global. In 2010 alone there were twenty-seven zombie films produced, ranging from *Big Tits Zombie* in Japan, to *Santa Claus vs the Zombies* in the USA. The zombie is now a staple enemy in video games – the predictable, dumb target who multiply the more of them you kill. There are zombie weekend conventions; zombie 'walks', where people cover themselves in gore to raise money for charity. The zombie has become a trope: a narrative framework understood by all, whose rules and conventions allow you to hang any other ideas inside it: so we get movies such as *Kung Fu Zombie, Biker Zombies from Detroit, La Cage aux Zombies* and *World War Z*.

Why are we collectively investing such a huge amount of concentration, emotion and mental energy into the zombie? What are we trying to say to and about ourselves?

Human cultures have always constructed myths and legends about undead beings or semi-humans, usually as a metaphor for some deep-felt human need – but the zombie is unique. Zombies are not vampires. The relationship between vampire and victim is a metaphor for illicit sexual attraction, plus you can reason with a vampire. Zombies are not ghosts. The metaphor behind the ghost story is grief, and ghosts can't kill you. Zombies are not werewolves: the werewolf is a metaphor for mental illness, or sociopathic violence – and becoming one is temporary, while becoming a zombie is irreversible.

Compared to the traditional monsters of Western folklore, the zombie has a superpower that sets it in a class of its own: it is self-replicating. One werewolf is not going to decimate London; one vampire will not depopulate Transylvania. One zombie, however, can – through an exponential process of killing or infection – take down an entire society.

So what is the real, deep fear that the zombie metaphor plays on?

The most likely answer is: the fear that we are about to lose what makes us human – our rationality, our capacity to discern truth from lies, our ability to see other human beings as fellow species members, with rights equal to our own. Our agency. Our freedom.

Such fears are rational. We are facing the biggest attack on humanism since it was formulated in the days of Shakespeare and Galileo. Humanism was central to Western ideas of civilization, to scientific thought and to concepts of social progress for more than 400 years. But since the late twentieth century, opposition to humanism has been building from several directions at once.

The strategic threat is from technology. It is possible that within this century artificial intelligence will attain a level of sophistication that exceeds the capabilities of all human brains put together. At the same time bio-engineering has advanced to the point where one-off modifications to individuals and – if the taboos on it were lifted – irreversible changes to humanity's gene pool are possible. Belief in these possibilities is fuelling a strong anti-humanism among those thinking about the future: a defeatism about the value of human individuality; a conviction that *Homo sapiens* is a species destined to be eclipsed.

Second, developments in neuroscience and information theory have strengthened the belief that our behaviour is inescapably determined; that our brains are just biological machines, 'programmed' by their DNA and modified only by their physical environment, within a universe which itself now looks more and more like the product of a giant 'computer'. Though both propositions are disputed within science itself, the airport bookstands of the world are groaning with bestsellers that ignore the nuances and convey the straight message: we are already automata incapable of freedom.

Third, there is a simple demographic fact: the majority of the earth's population now lives in countries where the cultural concepts underpinning humanism are weak. When the Universal Declaration of Human Rights was signed in 1949, there were 2.4 billion people on the planet, a quarter of them living in developed, democratic countries with social elites shaped by the traditions of the Enlightenment. Today there are 7.5 billion – and the majority live outside stable democratic systems, in societies where human rights are denied. Worse still, the official ideologies of these states are thoroughly anti-humanist. This includes the mixture of Confucianism and accountancy that is taught as 'Marxism' in China, the Hindu chauvinism of the Modi administration in India and the Great Russian nationalism that animates Putin.

Last but not least, there is the attack on humanism carried out over the past four decades in the name of free-market economics. By coercing us into new routines, forcing us to adopt new attitudes and values simply to survive; by reducing us to two-dimensional economic beings, the economic model known as neoliberalism has broken down our behavioural and intellectual defences against the subsequent forms of anti-humanism that are now coming at us in the early twenty-first century.

The inflection point, crystallizing all these dangers and accelerating them, was Trump's presidential victory, and the global wave of right-wing populism he helped unleash.

Trump launched himself like a wrecking ball against the multilateral institutions on which the globalized free market relied: the UN Human Rights Council, the World Trade Organization, the European Union and NAFTA. By stigmatizing the media as 'fake news' and by injecting gesture and unpredictability into diplomacy and domestic politics, Trump was not only trying to dismantle the post-1989 world order. He was actively trying to create disorder.

In his response to the Charlottesville violence in 2017, Trump gave a clear green light to a new form of fascism in the USA. The alt-right attacks the whole idea of universal human rights; it relentlessly questions the validity of scientific thought; it denigrates the institutions dedicated to producing objective truth, like universities or the publicly regulated media.

Meanwhile, the very tools Trump used to wage war on liberal, democratic values in the USA were machines that suck the lifeblood out of human choice and reason: the algorithms that Facebook supplied to Cambridge Analytica, so that Trump and his Russian supporters could manipulate the opinions and voting behaviour of US voters.

If this new alliance of right-wing authoritarians and techno-literate fascists get their way, large numbers of people are going to become like that farmer from Tennessee: dead-eyed, unthinkingly obedient, lacking any sense of agency, their behaviour controlled by Facebook algorithms and their thoughts merely an echo of last night's Fox News. Political zombies.

At the core of the authoritarian right's agenda is an attack on the possibility of truth. The aim of Trump and his imitators is to produce in the minds of millions the conviction that nothing is true: that all news footage is doctored; all images of war and torture are Photoshopped; all terrorist attacks are 'false flag' operations by some deeper and unguessed intelligence agency; all victims of war and torture are 'crisis actors'.

They want us to believe that the rule of law represents an attack by the 'deep state' against the popular will; that the professional news media are 'enemies of the people'; that political opposition parties are 'saboteurs'. Autocrats like Vladimir Putin and Narendra Modi were already operating from the same playbook, with fewer obligations to democratic principles, but Trump took the approach mainstream. His success, during the first twenty-four months in office, has inspired copycat projects in Brazil, Hungary, Italy and beyond.

We are even now underestimating the seriousness of the catastrophe that's unfolding. This is not some short-term political cycle. It's a global attack on methods of thinking, science and evidence-based policymaking which go back to the early seventeenth century.

And it is also a crisis for the dominant mode of thinking on the left. As you scroll through the obscene claims of the internet trolls – that the latest ISIS terror attack was staged by the CIA, or that some mutilated Syrian child is a 'crisis actor' – always remember that the groundwork for the attack on rationality was laid by a left-wing academic current called postmodernism.

'A theory', wrote the physicist Hermann Weyl, is a set of ideas that allow you to 'jump over your own shadow', using words and numbers to represent what cannot be physically seen.[5] The postmodernists replied: 'How can you jump over your shadow when you no longer have one?'[6] Jean Baudrillard, who wrote these words in 1994, believed our willingness to live as capitalism dictates, to the rhythms of money and self-interest, had hollowed out our humanity. We had become mere expressions of economic forces, unable to cast a shadow onto the world, incapable of thinking beyond the reality presented to us by mass media.

The academic left had theorized human helplessness long before the right turned it into a project. What began in the 1950s as an explanation for working-class passivity has now coalesced into a growing academic and philosophical movement called post-humanism. It is an outright rationale for our slavery to machines and, at its most extreme, our voluntary extinction as a species. One of this book's aims is to put the post-humanism industry out of business.

To defend rationality you have to defend what it is based on: the proposal that experience plus accurate observation can create verifiable truth inside our brains.

When you trust your life to an airliner flying at 40,000 feet, you do so because you believe there is a real world, independent of your senses, whose laws the aircraft engineer has understood. However complex that world is, however full of randomness, to retreat from the belief in the 400-year-old scientific method that guides the aircraft engineer would be a seriously retrograde step.

To debunk the new religions of irrationalism and fatalism we have to return to a way of thinking that has become deeply unfashionable, which places humanity at the centre of its worldview – not machines, not nature, and not subgroups of human beings with differential rights – but all of us as a species.

After the Holocaust and the Second World War, humanism was the liferaft the survivors clung to. In the aftermath of Trump's shock victory, a new generation delved once again into the great humanist writers of the antifascist era: George Orwell, Primo Levi, Hannah Arendt and the rest. But once you get beyond the similarities, and the comforting soundbites, it's clear that theirs was a worldview at odds with the assumptions of modern progressive thought.

Humanism became unfashionable because of its association with white, Eurocentric culture, its justifications for colonial domination and its alignment with male power. In the 1960s the black French psychiatrist Frantz Fanon called for a 'new humanism' devoid of the racism of the colonial past – but it didn't happen. Instead, from Vietnam to Iraq, devastating attacks on human life were carried out by politicians professing to be humanists. The French anthropologist

Claude Lévi-Strauss summed up the growing distaste for humanistic thinking when, in 1979, he claimed not only colonialism but fascism and its extermination camps were the 'natural continuation' of humanism as it had been practised for centuries.[7]

Then, towards the end of the twentieth century, neuroscience, genetics and anthropology all made claims that seemed to undermine earlier scientific assertions about what makes humanity unique. Meanwhile, some deep-green environmentalists concluded it would be better for the planet if we did not exist, while some radical supporters of animal liberation added: the sooner the better.[8]

The defence of rationality and science can succeed only if we return to a different form of humanism than the one espoused by Arendt, Primo Levi and their generation. There is, arising out of the same traditions of rationality and Enlightenment, an alternative and more radical form of humanism whose aim is complete liberation – including liberation from the identities imposed on us by poverty, racism and sexism.

Only one thinker in the humanist tradition combined *realism* – the idea that the world exists beyond our senses – with a definition of human nature that can withstand twenty-first-century theories of cognition and artificial intelligence. His name was Karl Marx. Despite all the flaws in his theories and all the crimes committed in his name, Marx was the only great philosopher who, had he been alive, would have gone masked up on that protest in Washington DC. He would have understood what it signified, too: Day Zero in the struggle to rekindle hope.

2

A General Theory of Trump

'Globalisation is dead. The American superpower will die.'[1] That's what I wrote in a column filed for the *Guardian* two hours after Trump declared victory. He had won, I suggested, 'because millions of middle class and educated US citizens reached into their soul and found there, after all its conceits were stripped away, a grinning white supremacist. Plus untapped reserves of misogyny.'

It was perhaps an extreme thing to write at a time when mainstream opinion writers were saying his victory had been an accident, the result of Clinton's campaign mistakes in four swing states, and would soon be remedied by Trump being smothered within the great federal bureaucracy and hogtied by the rule of law.

But Trump's victory was part of a pattern. This was the third tsunami to hit the liberal political centre in eighteen months. In June 2015 the people of Greece had voted to defy the EU, despite being held to ransom by the closure of their banking system. In June 2016 a clear majority of British voters opted for Brexit. And now, in November the same year, there was Trump.

I'd been warning since the 2008 financial crisis that, unless we ditched free-market economics, a major country would exit the multilateral system based on rules and common standards, and globalization itself would begin to die. The *Financial Times* called these warnings 'irritatingly shrill'.[2] Not shrill enough, as it turned out.

Trump's victory was not just an event in the political and economic history of the world, big enough though that is. It was a tear in the intellectual fabric of the world that, even now, most people have failed to understand.

Whether Trump is indicted, impeached or simply incapacitated through an overdose of cheeseburgers, his victory has irreversibly changed the world we live in. He declared war on the rules-based global system, started a trade war with China, pulled America out of the Paris climate change accord, destroyed the 2013 Iran nuclear deal, legitimized far-right violence, incited violence against the media, and brought organized lying into the mainstream of both politics and diplomacy.

His 'America First' strategy was not only about boosting US jobs and growth at the expense of China and Mexico, it was an attempt to shatter the existing global power structure and remake it, with America and Putin's Russia as co-beneficiaries. His tactics have included threatening North Korea with pre-emptive nuclear war, and putting migrant toddlers behind wire fences separating them from their parents. And, to date, he has succeeded.

To achieve the new order, the method Trump adopted was chaos: the outrageous statement followed by denial; the communiqué signed and then cancelled by mid-air Tweet; diplomacy conducted without diplomats, advisers, written records or accountability.

To orient ourselves amid this chaos we need a theory that explains how the new right-wing authoritarianism developed, who benefits from it and what it is aiming to achieve. That is exactly what most liberal-minded people did not have on the night of Trump's victory. They understood that this monstrosity signalled the potential end of liberal politics and of an orderly global system, but they could not comprehend it was the liberal order itself that had created Trump and empowered the activists who put him into the White House.

Even once we understand Trump we will only possess a theory of the wrecking ball. To complete the picture we will need to survey the fragile structures it has begun to wreck. These, it turns out, include not only the economic architecture of the world but the ideologies of liberalism, globalism and universal rights.

These ideas have become so fragile because they grafted themselves onto an economic structure that could not survive. During the thirty-year rise and fall of the economic model known as neoliberalism, much of its thought-architecture was expressed through performances and rituals that did not require inner belief. By the end, just as with the Soviet Union before it collapsed, people were going through the motions but knew in their hearts the whole thing was bullshit.

To re-establish order and predictability in the world, we need to restore what the neoliberal era stripped out of it: the three-dimensional human being with a belief in restraint, kindness, mutual obligation and democracy; an army of individuals who can think independently and who mean what they say. As you can imagine, this won't be easy.

Trump declared his presidential run on 16 June 2015 from a podium inside Trump Tower. In a rambling and apparently unscripted speech he outlined the key planks of his platform. He attacked Mexican immigrants, saying: 'They're bringing drugs. They're bringing crime. They're rapists. And some, I assume, are good people.'[3] He promised to 'make America great again' by forcing the US corporate elite to move jobs back onshore and through punitive trade sanctions against China and Mexico. He would reverse US foreign policy in the Middle East, isolating Iran and backing Saudi Arabia. He would repeal Obamacare, which had brought 20 million of America's poor into the healthcare system; he would spend billions on upgrading America's decrepit infrastructure while at the same time (and miraculously) reducing the national debt.

The establishment laughed. Anti-racists went predictably and justifiably nuts. He polled just 6.5 per cent among Republican voters. But within six weeks Trump was scoring 20 per cent: double the ratings of his closest rival, Jeb Bush, and leaving a long tail of whey-faced Christian fundamentalists far behind.[4] Few understood it then, but Trump – through his racist, misogynist, economic nationalist and anti-elite narrative – had created a populist bandwagon more effective than all the other populists, and unmatchable by the establishment candidates.

If we had 20:20 hindsight, the question we should have asked as Trump gained momentum is: what fraction of the rich and powerful will move behind him? But at the time such questions seemed pointless. Because free-market capitalism in the USA had produced a political monoculture in which the very idea that different sections of the elite could use politics to fight each other seemed to belong to the days of sepia photographs. The norm for thirty years had been a socially liberal business elite oriented to finance, global corporations, carbon extraction and tech monopolies. Their general preference was for a government of the centre right but ultimately the party-political divide didn't matter. Most big corporations donated to both parties.

Sure, there were by 2015 tens of thousands of ruined small business people and laid-off workers in the right-wing Tea Party movement, clamouring for an end to globalization, human rights and immigration. But their agenda was so contrary to the interests of the corporate elite that it could find support only among cranky individuals such as Charles and David Koch, prepared to pour $400 million down the drain of libertarian lost causes.

This in turn shaped the accepted wisdom among the pollsters. In April 2016 I sat through a briefing by pro-Clinton analyst Stan Greenberg, in which he assured the *Guardian*'s political journalists that the coming election was 'edging towards an earthquake' that would destroy the Republicans and put Hillary Clinton into power. The reason was that a 'new American majority' comprising black people, Hispanics, millennials and single women now made up 54 per cent of the electorate and rising. That made it impossible for the Republicans to win on a programme of social conservatism. Republican right-wing activists weren't even trying to win the election, he told us: they just wanted to punish the Republican mainstream for failing to stop Obama.[5]

Trump won the nomination because he created, first, a new kind of conservative populist movement. With it, he opened up a split within the US ruling class over where its material interests lie, both in geopolitics and economics. And with these two forces he created what Hannah Arendt had labelled a 'temporary alliance between the mob and the elite'. Its aim was the destruction of an economic and political order that had been presented as both perfect and permanent.

In 2012 I attended a Tea Party meeting in Phoenix, Arizona. It was a collection of pleasant, analogue-era cranks. Before we went in, I gave my colleagues a team talk about respecting such people's views. At the end people queued up to hand me files, folders and CDs wrapped with handwritten notes. There was a large file on the Obama birth controversy; a well-researched timeline of the Benghazi fiasco, where four US personnel had just been killed; plus the usual stuff debunking climate change. I took the whole pile of CDs, files and leaflets detailing their nutty obsessions and made my cameraman film me dumping them in a bin. Here's why.

At the start, I'd taken them seriously. In 2008 I reported on the mass mobilization of right-wing voters that derailed the Bush administration's $780 billion bank bailout in Congress. While others wrote them off as 'astroturf' – fake grass roots – I treated them as a genuine force, motivated by justifiable grievances over the way Wall Street made ordinary people pay for the financial crisis. After that, I'd watched with growing fascination as the Tea Party colonized the Republican apparatus from below. I'd stood in their rallies, enduring their scowls, because I knew the existing order could not last and I wanted to understand what might replace it.

But by 2012 it looked like they'd hit a dead end – an impression shared by many people in that Phoenix meeting. Mitt Romney, a moderate, was the Republican presidential candidate. As a result, most said they would refuse to vote. True, his running mate, Paul Ryan, had tabled an alternative budget calling for tax cuts, cuts to health and welfare programmes and a shrunken state. But the Tea Party was never just about economics. It was also a revolt against modern life by evangelical Christians; a revolt against women's liberation by misogynistic men; a revolt against immigration, gay rights and diversity; and above all a revolt against President Obama by those who could not stand the colour of his skin.

From Romney's defeat in November 2012 to the moment Trump descended his golden escalator in June 2015, the Tea Party would remain trapped inside the political ghetto I'd seen in Phoenix. Because alongside sacred America there is always profane America. In some states, along mile after mile of freeway, you see only adverts for roadside porn cinemas, liquor stores and the Confederate flag. Here the Jesus brigade could never become a popular movement. Their morals would not allow them to mix with the kind of people who sit glue-eyed on the slot machines at Trump's casinos or leering at the waitresses in the Hooters fast-food chain.

The Evangelicals were insistently nice people – even while waving plastic foetuses in the faces of frightened women outside abortion clinics. They had moral limits. But that was the problem Trump solved for the American right: he brought in the not-nice people, the amoralists and the self-described 'shitposters' of the online right.

In every Hollywood movie there is a text and a subtext. The subtext of the movie – which is never spoken – is what sends individuals out of the cinema prepared to join wars, save the planet or get divorced. Trump, like all demagogues, is a natural at manipulating text-vs-subtext.

The 'text' of the Trump campaign was Trump's life itself: a story of rags to riches. The riches were gained through speculative property investments and extensive business contacts with Russian oligarchs and Gulf oil sheikhs in an industry rife with organized crime. David Cay Johnston, a Pulitzer-Prizewinning journalist, writes that 'Trump's career has benefited from a decades-long and largely successful effort to limit and deflect law enforcement investigations into his dealings with top mobsters, organized crime associates, labor fixers, corrupt union leaders, con artists and even a one-time drug trafficker.'[6] By picking Trump to run for president, the Republican Party created a new and shocking subtext: the rich no longer have to even look clean to run America.

Once the campaign started, Trump inserted a second, equally shocking subtext into public life, about the irrelevance of facts. In July 2015 he insulted his opponent, Senator John McCain, saying: 'He's not a war hero. He's a war hero because he was captured. I like people who weren't captured.'[7]

When the remark provoked outrage, Trump denied he had ever said these words. The insult, its viral repetition across social media and then the flat denial told a story between the lines that would recur many times later: nothing Trump says is meant literally, nor should be taken seriously. Nor should any of Trump's utterances be held up against normal standards of truth or decency. This demonstration of blatant lying took Trump out of the league of previous US presidents and into the league inhabited by the standout kleptocrats of the twenty-first century: Russia's Putin, Turkey's Erdoğan, Hungary's Orban and Israel's Netanyahu.

A third layer of subtext was written at Trump's rallies. In the Tea Party movement, in front of the cameras at least, they usually tried to restrain outright bigotry. Trump thrust this nicety aside, saying to the racists, sexists and Islamophobes: go ahead and vocalize all the hate inside you. The rallies brought together a mixture of born-again Christians, amoralists from the alt-right movement and porn-addicted right-wing bigots – and created an atmosphere in which they could all yell the word 'cunt' every time he mentioned Hillary Clinton.

Trump is no fascist; nor were most of those at his rallies. Yet Trump played on a dynamic between speaker and crowd that was first theorized by the German sociologist Erich Fromm during the rise of Hitler. 'Psychologically,' wrote Fromm in 1941, people's readiness to submit to fascism 'seems to be due mainly to a state of inner tiredness and resignation', which he said was 'characteristic of the individual in the present era, even in democratic countries'.[8] Where this 'inner tiredness and resignation' comes from, in the richest economy in the world and a society buzzing with cultural creativity, is one of the most fundamental problems those trying to resist the new right have to confront.

Trump understood that tired people don't want logic or principles; and they don't want the kind of freedom that the libertarian right offers. In fact they fear freedom. What they want is a leader who rises above logic and truth and tells them all their inner prejudices are right. There is no mystery as to why the people at the rallies bought Trump's offer. But why did part of the elite buy it, and what do they want to achieve?

For the first months of the 2016 primaries, the money that would put Trump in the White House was invested in the ultra-right conservative Ted Cruz. Hedge fund boss Robert Mercer – who would become Trump's biggest donor – had given him $11 million, while four members of the Wilks fracking dynasty had handed Cruz $15 million between them. Fronting the Cruz SuperPAC was Kellyanne Conway, later Trump's presidential counsellor.

But the Cruz campaign faltered and Trump's took off. When Cruz pulled out in May 2016, Mercer's group effectively engineered a reverse takeover of the Trump campaign. By August, Steve Bannon – into whose far-right news outlet, Breitbart, Mercer had already pumped $10 million – was installed as campaign chairman and Conway as manager.

Meanwhile, a niche group of more traditional right-wing business leaders came out for the Trump project. They included casino magnate Sheldon Adelson; Carl Icahn, a property boss and asset stripper; and Wilbur Ross – another asset stripper who together with Icahn had helped save Trump's casino business in the 1980s. These were property and casino guys – sharks from the same shiver as Trump. Alongside them came a few libertarian tech billionaires, notably PayPal founder Peter Thiel, who had declared in 2009 that 'I no longer believe democracy and freedom are compatible.'[9]

The Koch brothers, the most prominent elite businessmen associated with the Tea Party, kept their distance from Trump on ideological grounds. But they unleashed millions into Republican Congressional campaigns, mobilized their army of paid canvassers and placed key people into the Trump team, notably Indiana governor Mike Pence. The Kochs had bankrolled Pence as he turned Indiana into a laboratory for free-market cruelty – now they would make him vice president.

However, even as Trump attracted more elite support, the bulk of billionaire money was going to Clinton. Trump had the casino guys, big oil and big tobacco. But Clinton had most of Silicon Valley, most of Hollywood, most of Wall Street and most of the S&P 500. Even the heiress to the union-busting Walmart empire backed Hillary.

Once Trump won, of course, many of these business people fell over their own shoes to congratulate him, join his advisory boards and take part in the bonanza of deregulation he offered. But those given direct power were still drawn from the tiny right-wing circle that had driven the project. Betsy DeVos, the school privatizer, was put in charge of schools. Wilbur Ross, at the age of seventy-nine, was made commerce secretary. Rex Tillerson, whose Exxon Mobil had funded climate science denial, became secretary of state. Robert Mercer's daughter Rebekah got an executive role, while the Trump business empire itself was represented by Jared Kushner, the president's son-in-law.

So to describe this as a 'corporate takeover' of US politics, in the words of left-wing Canadian writer and thinker Naomi Klein, is too simplistic.[10] It was a takeover by a minority fraction of the business elite, its centre of gravity sitting squarely in the world of private companies untroubled by stock market scrutiny, and with overlapping

aims: massive deregulation, a trade war on behalf of domestic industries and a radically shrunken state. From Adelson to the Uber
founder Travis Kalanick, these were executives prepared to hijack the
state to deliver favours, contracts and privatized assets to their own
businesses – rather than play the official game of stock market-listed
companies operating on a level playing field.

Since the early 1990s this official game had delivered something
close to what Karl Marx described as 'capitalist communism'.[11] It
works like this. Through the quarterly financial disclosure required
of companies listed on Wall Street, the average profit margin in a
business sector becomes clear and predictable, especially if the sector
is mature. Then the finance system begins to work as a sharing mechanism, in which everybody with capital can participate. When
America was an industrial superpower, financial profits made up just
15 per cent of the profits. By the mid-2000s finance was generating
40 per cent of profits.[12] So long as everybody could dip into the financial
cookie jar, and the state was seen to crack down on those stealing
from it – as in the Enron case and the Wall Street analyst scandal –
few rich people questioned the dominance of finance.

At the same time, corporations understood that their common
interests were being represented globally by the American state. Since
1979 the USA had tirelessly imposed deregulation and free trade
onto less powerful countries, and relentlessly borrowed money from
them on terms rigged in favour of itself. Globalization worked in the
interest of US business and the US government had used its power to
enforce it on the world – even if that also meant the impoverishment
of America's traditional industrial communities. That was the deal.

Then came the 2008 crisis. As the long-term costs of stabilization
became clear – permanent state intervention, banking regulation and a
giant debt – it tore apart political consent among America's rich, both
for globalization and for the 'level playing field' between firms within
the USA, mediated by the finance system. With growth stagnating,
with climate regulations placing new burdens on carbon-heavy businesses, and with bank profits suppressed by increased regulation, a
fraction within US capitalism broke with the political consensus.

Instead of globalization they wanted a form of 'national neoliberalism': free-market economics pursued not as a benign global strategy

for all rich people in the world, but to enrich the US elite alone, if need be at the expense of their foreign counterparts. As for the cookie jar of finance, they wanted the right to dip in first and often, to the detriment of everyone else. Trump was not their chosen candidate: Cruz was. But Cruz was a dud and Trump was not.

Trump's mesmerizing incompetence and verbal brutality have framed the political situation so completely that, for many people, he 'is' the crisis. But he was, in a way, simply the accidental front-man.

In February 2016 the NFL staged the last Superbowl of the liberal era. The ad breaks featured the familiar mix of foreign automobiles and American carbohydrates. The half-time show starred Beyoncé, with a dance troupe dressed as Black Panthers from 1968: a clear reference to the Black Lives Matter movement, it was meant to symbolize the contrast between the bad old days and now. The USA is a multiethnic democracy, with a recovering economy and the political maturity to stop its police force shooting black people at will. That was the subtext.

By this point, the recovery that had begun in spring 2009 had created 17 million new jobs.[13] The Dow, which had slumped below 7,000 in March 2009, was now above 17,000 and rising. GDP stood at $18 trillion, 4 trillion higher than it had been at the start of the recovery. On top of that, the USA was on the point of signing major trade deals – the Trans-Pacific Partnership (TPP) and the Transatlantic Trade and Investment Partnership (TTIP) – each designed to create an even bigger market for American goods and services.

Why would a section of the elite put all this at risk for the sake of economic nationalism? Conversely, why did the corporations in whose interest it was for globalization and bipartisan politics to continue fail to fight for a clear alternative? To answer the first question we need to look at the coalition of interests represented by three very different Trump backers: Robert Mercer, the Koch brothers and Stephen Bannon.

Mercer has never made a public speech. But thanks to lawsuits and briefings by ex-employees we know some of what he is alleged to think: that the lethality of nuclear weapons is over-hyped; that radiation made Hiroshima's survivors healthier; that black people were better off before the Civil Rights Act; that climate change will improve life on earth. Mercer is said to have told colleagues that the state 'makes the strong people weak by taking their money away, through taxes'.[14]

Mercer is a computational linguist who used his expertise in data analysis to build a hedge fund that has generated $55 billion in profits. It has no investors other than the people who work for it – a couple of hundred quantitative analysts, known as 'quants'. They pay no taxes, as their profits are invested in a pension fund. The profits are generated by advanced mathematics – but nobody knows how, because Renaissance Technologies (RenTec) is what financiers call a 'black box': it is a machine that works without explanation.

RenTec makes profit by exploiting its ability to see patterns in the numbers generated in financial markets, which in turn are generated by trillions of transactions in the real world. Its analysts found, for example, that global markets perform better on sunny days. So they built a model to exploit that. The return on money invested in a good year like 2008 was 98 per cent; and across all the years of crisis between then and now, Mercer's Medallion fund has never made profits of less than 28 per cent.[15]

All businesses have a narrow interest, which they have to pursue at the same time as compromising with the wider needs of capitalism. What is RenTec's material interest? Well, if Wall Street is like a ranch where the cows are ordinary businesses, RenTec is like a ranch where the cows are Wall Street plus every financial market in the world. It can 'farm' the financial profits of other companies and investment banks because it owns a machine that can think faster than everybody else.

As long as there is a market, and unpredictability, and some capital to invest, real-world factors such as the tax rate, US trade policy or the quality of public healthcare simply do not matter to a company like this. It has technically no social obligations or interests. RenTec's ultimate material interest lies in knowing more than everybody else and there being enough unpredictability for this to matter. Its ideal environment, therefore, is chaos.

Koch Industries is a more traditional antisocial business empire. The wealth of its owners, Charles and David Koch, was built on oil and industrial processing, and maintained through the usual means: avoiding taxes.[16] Their narrow material interest is more traditional: the removal of obstacles to profitability – like the minimum wage, corporate taxation, publicly owned land, environmental protection laws and carbon emission caps, the public healthcare safety net and the tax-funded pension system. They want it all swept away.

However, it would be wrong to see the Kochs merely as the spearhead of a US corporate deregulation drive. Their goal is effectively a form of capitalism without government. When, back in 1980, David Koch stood as the Libertarian Party's candidate for vice president, he called for the abolition of the federal authorities regulating air transport, electoral law, environmental protection, food standards and the energy grid, together with all state provision of education, basic healthcare and retirement pensions.

That is not the project of traditional 'small-state' conservatism: it is capitalism without a state, in which the most powerful are left free to accumulate wealth and coercive power, buy votes, poison the waterways and exploit the old, sick and poor, for whom there is no safety net whatsoever. The Kochs' conservative critics called it, at the time, 'anarcho-totalitarianism'. A shorter word for the project might, again, be: chaos.

If you're wondering how the libertarianism of the Kochs fits in with the techno-mysticism of Mercer and the avaricious ego of Donald Trump, the missing link is Steve Bannon. A former Goldman Sachs executive turned Breitbart News executive, Bannon is an economic nationalist. In pursuit of American nationalism he has, through Breitbart, promoted a panoply of racist, Islamophobic and white supremacist views, titling a whole strand of Breitbart's output 'Black Crime'.

However, what defines Bannon's project beyond these ordinary prejudices is a theory of history known as the Fourth Turning. According to its authors Neil Howe and William Strauss, political systems typically rise and fall in four phases: in the first there is exhilaration and strong identification with the state; then comes an 'awakening', during which people connect with deeper principles; then comes disorder, where loyalty to institutions falls apart. Finally, in the 'fourth turning' there is a systemic crisis – a revolution – triggered by a survival-level threat. If you see the post-1945 boom as phase one, 1968 as the awakening and the post-Nixon turmoil as the disorder, you might conclude that we've been waiting a heck of a long time for phase four. But, in Howe's words: 'If history does not produce such an urgent threat, Fourth Turning leaders will invariably find one – and may even fabricate one – to mobilize collective action.'[17]

Finding such urgent threats has, effectively, been Bannon's mission since the mid-2000s. The threats he has invoked include jihadi terrorism, China (with whom Bannon has predicted war), Mexican immigration, 'black crime' and the US national debt. But since none of these threats has actually mobilized masses of Americans to stage a revolution, there was always the ultimate option – as Howe suggested – to fabricate something. What Bannon fabricated was the chaos of the Trump presidency. Once it started, it didn't need Bannon's guiding hand for long: Bannon got himself kicked out of the White House and switched to amplifying the impact of Trump's chaos strategy across the Western world, by attempting to build an alliance of ethno-nationalists committed to destroying the European Union.

For Bannon and his alt-right followers, what we're heading for is an event paralleling the run-up to the American Civil War of 1861 – except, this time, with global consequences. In this scenario, America's long-simmering culture war becomes a low-level armed conflict with parallels in the 'Bleeding Kansas' crisis of the 1850s; meanwhile, external threats are invoked to justify suspending aspects of the rule of law; finally Trump or his replacement starts a major conventional armed conflict. The resulting destruction of the post-1945 world order then re-sets the mass psychology of the USA, giving legitimacy to a new, authoritarian, nationalist ruling elite.

From three separate sources, then – a hedge-fund strategy, ultra-libertarianism and a theory of history lifted from the airport bookshelves – Trump's backers converged on this strategy of chaos.

Mercer, Bannon and the Kochs each contributed key parts of the machinery that would put Trump in power. The Kochs' machine was an alternative party – a shadowy network of think-tanks, canvassers, phone banks, petitions and vote-suppression projects. Bannon's machine – Breitbart – produced false or biased news, often based on the output of neo-Nazi and white supremacist websites – to be picked up as 'talking points' by Fox News, echoed by the president and swallowed by his followers.

Mercer's machine was Cambridge Analytica (CA), a data analysis company which had scraped from the digital records of every American voter 5,000 pieces of information that could, it claimed, predict their voting behaviour better than any other model. Cambridge Analytica allocated thirteen staff to the Trump campaign to help target radio, TV and internet advertising not just towards states, postcodes or demographic groups, but towards individuals. Based on CA's real time data, if social media showed a spike of discussions about immigration in a swing state, Trump could immediately stage a speech about it based on the intelligence.

Together these three machines formed an efficient production line for precisely targeted lies.[18]

What made millions of people vote for a man who promised to dismantle the state and blow up the international order? In the days after Trump's victory, just as with Brexit, there was an outpouring of journalistic garbage about the 'white working class' and its economic deprivation. We were asked to believe that the people who had put Trump in power were primarily low-income and that their main grievances were stagnant wages, inequality and other impacts of globalization. All the evidence shows this is wrong.[19]

A survey for the Public Religion Research Institute (PRRI) found anti-immigrant views and cultural fears were 'more powerful factors than economic concerns in predicting support for Trump among white working-class voters'. It also showed that working-class people who'd experienced economic hardship were more likely to support Clinton than Trump. What predicted support for Trump was not hardship but economic fatalism – where respondents identified a college degree as a 'gamble' rather than a smart career move.[20]

The pollster Gallup, which crunched the data on 125,000 voters, found Trump-supporting households were earning on average nearly $5,000 a year more than Clinton-supporting households. The economic factors that correlated with support for Trump were above-average disability and early death; or living in a place where there is poor social mobility from one generation to the next. Basically, deadbeat towns cut off from the global economy. Conversely, Gallup found that the higher the number of manufacturing jobs in an area, the lower the support for Donald Trump.

Nor was it the presence of migrants or black people in an area that drove people to Trump. 'The racial and ethnic isolation of whites at the zip code level,' concluded Gallup, 'is one of the strongest predictors of Trump support.' All along the line it is the absence of contact with global modernity that predicts support for Trump, just as it did for British voters supporting Leave in the Brexit referendum.[21]

Researchers based at the University of Massachusetts confirmed these results. The election had exposed a divide between educated and uneducated whites, they said, but 'most of the divide appears to be the result of racism and sexism in the electorate'.[22] Significantly, this study also found that, among men, being an active misogynist, as opposed to just a patronizing sexist, was as strong a predictor of supporting Trump as overt racism. Being poor came nowhere close.

So, while stagnating incomes and collapsing real wealth set the backdrop, Trump supporters were primarily waging a race and gender war, not a class war. Trump won, in other words, because large numbers of Americans harboured untapped and unchallenged reserves of racism, cruelty and misogyny. And identifying racism and misogyny as the key factors driving white voters to Trump allows us to understand what links their project to that of billionaires like Bannon, Mercer and the Kochs.

To use Arendt's terminology, both the 'elite and the mob' were attached to theories that no longer explained the world. So the world had to be reordered to fit what was in their brains.

For the Trump-supporting elite, their basic theory of self-correcting markets and the small state failed back in 2008. For the racists, covert theories of white supremacy going back to the days of slavery had long been challenged by the economic advance of blacks, Hispanics and other migrants. But their surviving assumption – that America would always reward whites with decent jobs, respect and cultural supremacy – was undermined after 2008, both by the economic crisis and by the Obama presidency. In repeated polling studies, it is the feeling of 'white vulnerability' and racial resentment that drives people to support Trump: their racial anxiety creates the economic anxiety, not the other way round.[23]

With misogyny, the roots go even deeper: women's oppression is traceable throughout the entire 40,000 years of recorded human history. Yet in the fifty years following the rollout of the contraceptive pill, developed-world society has experienced what Federal Reserve chief Janet Yellen called a 'reproductive shock'. The results do not come anywhere close to women's liberation – but increased access to jobs, greater sexual freedom and improved legal rights have changed the world for American women in just two generations. The basic assumption behind misogyny, that women are destined to stick to their biologically determined role as child-bearers and unpaid domestic workers, has been blown to smithereens.

And here is where we begin to understand the historic nature of Trump's victory.

Each of these ideologies – the nationalist neoliberalism of Trump, the white supremacy and misogyny of his supporters – was founded on a biological claim about human nature: that blacks are inferior to whites; that women are born to serve men and reproduce; that everybody on earth is genetically inclined to compete, to maximize their personal wealth and to stab each other in the back in order to do so.

But the forms of capitalism upon which these ideas were based no longer existed. The Doris Day era of racial segregation and female obedience had evaporated in the 1960s; the market fundamentalist paradise became impossible after 2008. What the right-wing section of the elite and their followers among the working class share is a desire to restore society to its supposedly 'natural' order. And for that they needed what Hannah Arendt once described as 'access to history': the ability to make changes in reality so that it would once again fit their theories of biological inequality.

Conservative claims about human nature are, of course, centuries old. From the mid-1960s onwards it seemed that social liberalism and scientific rationalism might blow them away. Instead, after the crisis of 2008, the opposite has happened.

As to the question of why the liberal, 'coastal' majority of the US elite either didn't understand the danger or defend the old corporate model harder, this goes deeper than political incompetence. The paralysis of institutions like the *New York Times* in the face of Trump and the complacency of the Clinton campaign also reflect something structural within the model that failed in 2008.

In this model – an economy dominated by finance topped by a stable, bipartisan political system – corporate leadership became depoliticized. It was commonplace in the era of globalization to hear that corporations 'dictate' to national states. But if that's the case, they did so via strictly technocratic methods: political donations, lobbying, the tame think-tank and the box at the opera house. On the other side of the conversation they expected to find – because they had created it – a technocratic state: a civil service governed by rules and laws, a relatively level playing field when it comes to competition, and a meritocracy when it comes to leadership. It was not necessary for senior managers of Boeing, Nissan, GE or Google to depict themselves as a liberal 'fraction' of US capital, because their project had relied on the absence of an opposing fraction, and indeed the subservience of the state to business as a whole.

If one segment of the business elite has become a show dog and the other an attack dog, it is a one-way fight. Anybody expecting the technocrats who run the global corporations to become counterattack dogs on behalf of democracy and human rights may have a long time to wait.

So Trump represents something bigger than a takeover of the federal government by one fraction of US capital devoted to protectionism and the small state. He represents the triumph of a reactionary theory of human nature in which inequality – of race, sex and economic status – is determined by our genes. This, as we will see, is the problem that's going to be hardest to overcome, because it is deeply rooted in the economic practice of the past thirty years.

That 'the Russians did it' was a comforting illusion for Clinton-supporting liberals in the aftermath of defeat. As evidence continues to emerge, it is clear that the Russian state made a major effort to help Trump achieve power, to drive the popular bigotry that sustained him, to feed intelligence to his campaign that had been obtained through hacking, and to place sympathetic people inside his team.

But it is clear that, in each case, the Kremlin was exploiting a systemic weakness of US capitalism itself.

The first weakness was the soft isolationism pursued under Obama. When he failed to respond firmly to Russia sending troops to Syria, to the Syrian chemical weapons attack in Aleppo, or to the Russian annexation of Crimea, Obama sent a signal about the future direction of the world. There may be sanctions against Russia now, Putin could assume, but in the long term there would be accommodation. The West would remain hospitable to Russian money and a sitting duck for Russian organized crime – whatever rules the Kremlin might break.

All of which created the climate in which Trump associate Paul Manafort could run, from inside the USA, a business promoting the interests of the Russian puppet government in Kiev.[24] It created the climate in which Russia Today, the Kremlin's propaganda channel, could pay former US general Mike Flynn $34,000, and in which Flynn could fail to disclose that money, even as he prepared to become Trump's national security adviser.[25] In the same climate Trump staffer George Papadopoulos could establish covert links with Russian agents offering to supply 'dirt' on Hillary Clinton. Meanwhile, Trump's son-in-law Jared Kushner could stage a meeting with a Russian intelligence asset, together with Manafort and Donald Trump Jr in Trump Tower, ostensibly to discuss the same idea.[26]

Security experts had warned for at least a decade that Russia was evolving a 'hybrid warfare' strategy, using a mixture of corruption, propaganda and organized crime alongside more traditional methods, to destabilize the West.[27] On current evidence it looks as if, once Trump's bandwagon began to roll, Russian intelligence stuck to his team like a humid day, and found numerous people on the US right prepared to wield influence on behalf of the Kremlin.

A second weakness they exploited was systemic financial secrecy designed to hide elite wealth from the tax authorities and help global finance evade regulation. This, indeed, is what allowed all the major intermediaries between Trump and Russia to hide their activities until after they gained office.

The same culture of secrecy allowed major businesses in the tech world to hand the Russians the means to intervene in the 2016 US presidential election. Facebook, Twitter and Google together provided the platform for fake accounts, bots and advertising manipulated by Russian intelligence to the tune of $100,000. Facebook in particular, whose algorithms are precisely tuned to reinforce the prejudices of its two billion users, was turned into a machine for spreading Russian lies.

Some 120 fake pages set up by Russian intelligence offshoots produced 80,000 posts, reaching up to 126 million people. On top of this the Russians spent tens of thousands of dollars on advertisements promoting 'distrust in political institutions and [to] spread confusion'. Facebook, which had willingly shut down the pages of Syrian human rights defenders, allowed Russian intelligence to play it like a pipe organ.[28]

Now let's do a thought experiment. Imagine you are a Russian intelligence boss: what is your analysis of US democracy's strategic weakness? At the root of everything is the same problem – deregulation and secrecy – which brought down the banking system in 2008. America's people are split by a culture war. Plus, parts of its elite actually have a material interest in creating chaos. Meanwhile, in the shape of Cambridge Analytica and Facebook (among others) they have created opinion control algorithms that allow their democracy to be manipulated by anyone with money. Though officially a rules-based democracy, the USA had become, after decades of free-market economics, a rules-free environment for anyone with technological or financial power. All Russia had to do was to use it.

Trump represents a triple catastrophe: a victory for racism and economic nationalism; a geopolitical slam-dunk for Vladimir Putin, triggering the breakup of the rules-based global order; and the first 'proof of concept' that corporate technology platforms can be used to shape the behaviour of a mature electorate. All this remains true despite the indictments, resignations, investigations and disarray attending Trump's first term in office. None of it will disappear automatically if he is indicted or defeated at the polls.

At the same time, Trump's victory dramatized a deeper crisis facing all advanced democracies. Even with strong economic growth, the system no longer delivers enough wellbeing and security to ordinary people to win their consent for it. Support for democracy and human rights is fading. Meanwhile, the secretive algorithms of the tech giants have become a deadly weapon against the very progressive values these firms are supposed to embody.

If the populations facing this threat were grouped into resilient organizations and had a strong sense of their own social power, the task for people like Putin, Erdoğan, Salvini and Trump would be harder. Their predecessors in the 1930s resorted to fascism because they had to smash an organized, politicized working class with a strong attachment to democratic rights, and a resilient liberal middle class inspired by the moral values of Christianity. That's what fascism was: the militarization of a lower-class mob to defeat the organized working class by force, take the state, merge it with the fascist militias and enforce rule by terror on behalf of big business.

This time round they probably don't need fascism. Solidarity has been atomized, our belief in collective action eroded, our sense of self hollowed out by the routines of market behaviour – and with that, so has the moral basis for liberalism. If you wanted to choose a moment to unleash an attack on democracy, reinforced by machine control of human behaviour, this would be it.

None of the forces that put Trump in power are invincible. History tells us that even billionaires can go to jail; that Russian despots can be overthrown. As for armed plebeian racism in America, it was defeated in 1865 – though only after five years of civil war.

The problem is, Trump was produced by a broken economic system and a geopolitical instability that are only going to intensify. Even if Trump were swept from office, we are now in a world where every four years a new, crazier, more vicious version of Trump is possible. A world in which the torchlit marches of the alt-right don't go away; in which the ideology of violent misogyny can spread from one generation of frustrated young men to the next.

To prepare ourselves for the blow-up that is coming we need a better understanding of what happened over the past thirty years – not just to the economy but to our collective human psyche, our sense of agency, our belief in reason. In 2008 we began to understand what damage neoliberalism had done to the economy; only in 2016 did we begin to see the damage it had done to our selves.

PART II

The Self

Only when a world order collapses does thinking about it begin.

Ulrich Beck[1]

3

Creating the Neoliberal Self

One of my earliest memories is of going to the Leigh Miners' Gala in the 1960s, when I was about five years old. Amid the tight throng of people in a field, there was a boxing ring in which a local slugger was taking on all-comers. One challenger had blood on his face, another a deranged smile: they were mostly drunk, their flesh raw from the fighting. What I remember most was my dad's hand sliding over my forehead to cover my eyes.

That scene took place at the height of the long post-war boom. Most of those in the crowd had experienced year after year of rising real wages. Many of the men, being miners, worked for the state. Their kids were educated by the state for free; healthcare was free; the water people drank, the energy they consumed and – for many – the homes they lived in were all provided by the state at low cost.

It was a world structured around an explicit deal between capital and labour. It feels like a lost civilization now, but a version of it was present throughout the industrialized world. If you want to understand why so many voters over the age of sixty yearn to go back to it, and why so much of the support for right-wing populism comes out of the ruins of it – from Northern France to Western Australia – you must understand that deal was unique, in what it delivered and the kind of person it created.

My grandfather went down Astley Green Colliery at the age of fourteen, at the outbreak of the First World War. My dad went down the same pit at the age of eighteen, in the final months of the Second World War. So, as far back as I can trace it, my paternal family tree goes: hat-maker, hat-maker, miner, miner, miner, economics editor. The post-war economic system, in short, delivered a lot more than an end to poverty and unemployment, and a bit of dignity at work. It delivered spectacular upward social mobility in your lifetime.

What replaced it has delivered a social catastrophe.

When I campaigned in Leigh for Labour, during the 2017 general election, what struck me was the large number of disabled and elderly people among the small crowd of activists who had turned up in the town square. Many were, as they readily explained, suffering from work-related disabilities or long-term mental illness. Most of them looked ten or fifteen years older than me. But when I looked closer at this rickety, grey regiment I realized they were my exact contemporaries.

In the 1960s a new glass and concrete office block had been built on Leigh's Victorian main street, signalling the arrival of white-collar work and the technocratic culture. Now it was a ruin. As we gazed up at its dust-caked, broken windows, a local councillor whispered to me: 'We're dealing with 10,000 cases of domestic violence a year.' The population is just 50,000.

'You can feel the despair, the absolute lack of hope and ambition, it's just been destroyed,' a sixty-year-old energy worker told me. A childhood friend who'd spent his working life down a coal mine said: 'At my old school the police are outside the gates at home time, looking for this or that suspect. There's organized crime: drugs, armed robbery. It's an industry – so many foot-soldiers on the streets selling drugs. If you send your kids to the shop there's a dealer waiting.'

That town had voted Labour since 1921, with only around 20 per cent of voters – skilled workers, managers and shopkeepers – habitually supporting the Conservatives. In the Brexit referendum of 2016 two-thirds voted to leave the European Union and, in addition to 20 per cent for the Conservatives, polls now registered a consistent 20 per cent supporting the xenophobic nationalist party UKIP.

The name for what happened to that town – and thousands of towns and suburbs like it all over the world – is neoliberalism. The elite doesn't want to talk about neoliberalism. They will tell you neoliberalism 'doesn't exist', or that the word is just a left-wing term of abuse. For them, the brutal economic logic imposed on places like Leigh in the 1980s 'just happened' – it doesn't need a name.

But we do need to talk about neoliberalism. Because in destroying the economic deal between capital and labour, it obliged millions of people to adopt a new self-image. Ways of thinking and behaving that would have seemed deviant to the people at that Miners' Gala became normalized over the past thirty years.

And now that neoliberalism is in crisis, these carefully ingrained behaviours, reflexes, thought patterns and self-images have also been thrown into crisis. What began in 2008 with the breakdown of the neoliberal economic system has triggered the breakdown of the neo-liberal self.

Neoliberalism is the specific global model of capitalism that began in 1979 and is currently falling apart. Though some countries have adopted free-market policies enthusiastically, others reluctantly, I am more interested in the totality: how all the different parts of the global system worked together – and then suddenly stopped working.

People who support neoliberalism often demand that its critics provide a definition of it. I could give you plenty of adequate definitions, the clearest being: competition forced into all aspects of society by a coercive state.[1] But the demand for definitions is a trap.

To understand complex, changing and uncertain things like economic systems, we need to train our minds to contemplate (a) the whole phenomenon and (b) the contradictions within it. We have to prepare for the fact that the appearance of things may differ from what's really going on beneath the surface – as with the banks in the run-up to 2008. We have to assume that all economic systems are temporary, and that their failure is often driven by the same factors that drove their success. This, for obvious reasons, is an uncomfortable way of thinking for the elite.

Instead of a definition, I want to outline a core set of relationships around which the neoliberal system's mutations, shocks and improvisations happen. There are three building blocks to any capitalist economy – land, labour and capital – and they produce money in the form of rent, wages and profit. Let's start by understanding how neoliberalism changed the relationship between these things.

During the state capitalist era (1945–79), the market was subordinate to the state. Labour and capital worked in partnership. As for rent, it was discouraged. When economists use the term 'rent' they don't just mean the rent on land or property, but any money extracted by cornering the supply of something – be it a cobalt mine, the fishing rights on a river or even the ability to raise capital itself. Rent does not create wealth. It merely distributes it from the people who produce wealth to those who own the rentable property – the 'rentier'. When he designed the state capitalist model, the economist John Maynard Keynes advocated the 'euthanasia of the rentier' – policies designed to drive rent-seekers out of the system.[2]

In the neoliberal era, by contrast, the state is subordinate to the market; indeed, the state's purpose becomes to sweep away all obstacles to the market and to force it into all aspects of life that remain non-commercial, from the supply of tapwater to arranging a date. Capital attacks labour, so profits rise as a share of GDP and the share of output being spent on wages falls. Meanwhile, 'rent' becomes a way of life. More and more profits flow to those who can create monopolies and set artificially high prices – whether these be software giants like Microsoft, social media giants like Facebook, investment banks like Lehman Brothers, or just the payday lenders in your town charging 1,000 per cent interest.

In its final phase, neoliberalism – which started out as a fight for free-market values – has become an unfree market rigged in favour of monopolists and speculators; rigged to protect the wealth of those who already have it; rigged to produce high inequality – a situation guaranteed by the elite's control of the state.

However, neoliberalism is not just the latest model of industrial capitalism. It is profoundly different from all previous models in three ways.

First, it is a model obliged to seek the destruction of organized labour rather than make a paternalistic bargain with it. This is as true in Shanghai as it is in Virginia. As a result, neoliberalism relentlessly disrupts the physical, social and institutional surroundings within which people have lived for generations.

Second, it is transnational. It creates a global market and globally distributed industries alongside control mechanisms that stand above nation states. As a result, for the first time in modern history nation states have been redesigned to act on behalf of a supra-national elite, whose wealth is primarily financial.

Third, neoliberalism was moulded around the rise of information technology, and information technology disrupts mechanisms that have been at the heart of capitalism for 250 years: the ability to keep prices significantly higher than production costs, and the ability to create new jobs for people whose jobs were taken by machines.[3]

I want to pause and contemplate the significance of the changes brought about by neoliberalism. Each involves a shift in the balance of power: away from those who work, away from national democracies and away from people who do not own technology companies. This power shift makes a new kind of disaster possible. If you invent automobiles, you also invent motorway pile-ups. If you invent a form of capitalism where power surges suddenly towards an unaccountable and technologically armed elite, with a penchant for class confrontation, it becomes easy to destroy the liberal, democratic and universalist ethos most people in the West thought was permanent.

The history of neoliberalism breaks into four clear phases: an upswing, from 1979 to 1989, when it was being imposed as a policy; a heyday, between 1989 and 2001, when it seemed to function automatically and went global; a manic period between the dotcom crash and the Lehman Brothers collapse; and the years between 2008 and 2016, when the cost of keeping the free-market model alive began to erode the geopolitical order of the world.

In each phase we see the consolidation of a set of economic relationships, assumptions, behaviours and ideologies that create a new image of the self in the minds of millions of people.

In the first phase, the countries driving the change were Britain and the USA. Both adopted economic policies that made the 1979–82 recession worse, in a way designed to destroy jobs, slash wages, shrink public services and above all erode the power of trade unions. Then they imposed these policies on other countries, using the IMF, a new global treaty on trade, direct political pressure and indirect pressure via the newly deregulated financial markets.

The new way of thinking was imposed on the electorate through punitive actions, much in the way a rogue dog-trainer uses violence. Every crack of the whip was meant to teach millions of people a lesson, not through newspaper articles or speeches, but through them seeing the results.

The first whip-crack was inflicted through monetary policy. In place of human outcomes – like employment or poverty – both Thatcher and Reagan focused economic policy onto achieving abstract mathematical goals, such as the money supply or inflation targets. The result was the rapid and severe destruction of entire industrial sectors. That taught us Lesson #1: in economic policy *humans no longer matter*.

To understand why the elite voluntarily destroyed entire towns and factories in the 1980s, you have to understand the social power of the people I'd seen at the Miners' Gala. By forming a movement of its own, and maintaining it for over a hundred years, the working class in the industrialized world had created a permanent counter-power, both to capital and the state. When he took me into that throng around the boxing ring and then covered my eyes, my dad was telling me a moral story. We have to coexist with the brutality of the industrial lifestyle, be part of it, learn to love its rhythms, smells and sounds – and at the same time we have to nurture the belief in something better.

Organized labour was, until the 1980s, the main humanizing force within capitalism, far exceeding philanthropy and religion in its material achievements. It brought us the weekend, the eight-hour day, the vote for those who don't own property, equal pay legislation for women – and it was the labour movements which, from Warsaw to Turin to Paris, took up arms to topple Nazism at the end of the Second World War.

The aim of neoliberal policy in the early 1980s was to inflict a slump so hard it would destroy the bargaining power of trade unions, the culture that incubated them, the values of solidarity they spread, the socialist ideals they nurtured and the workplaces they organized in. To break their will to resist, millions of skilled working-class people of my father's generation would be subjected to the very thing that had haunted their nightmares since childhood: the humiliation of poverty and long-term unemployment.

But even that was not enough. You also had to break people's belief in something better. You had to change their way of thinking.

The next crack of the whip was against France, which in 1981 had elected a socialist–communist coalition government, led by François Mitterrand. Mitterrand pledged to resist neoliberalism: he hired 200,000 civil servants, raised the minimum wage by 39 per cent and nationalized twelve industrial groups together with thirty-six banks.[4] In response, money equivalent to 2 per cent of French GDP left the country in the first three months. Three sharp devaluations of the franc against the German mark followed – and the final one, in March 1983, forced Mitterrand to abandon state-led growth and instead adopt austerity.

Mitterrand's government was forced, effectively, to occupy its own country on behalf of an outside power – the global financial markets.[5] Though it was Thatcher who said 'there is no alternative', it was the drama of France between 1981 and 1983 that drilled home Lesson #2: *left-wing alternatives to neoliberalism will always fail, because the financial markets will always sabotage them.*

The third lesson was taught through mass privatization programmes, either adopted voluntarily or, as in the case of Latin America, imposed by the IMF. Spain for example, sold or gave away thirty-four publicly held companies in the mid-1980s – almost always to foreign firms. In order to convince Volkswagen to buy the car maker SEAT, the Spanish government spent $1.5 billion writing off its debts, and $3.2 billion more absorbing losses and in state aid.[6] The workforce in the privatized firms was slashed by one third.[7]

The privatization process created a new group of people with a stake in neoliberalism's success: individuals who'd been given shares or allowed to buy them cheaply on privatization. They now had a new kind of logic in their heads: you, the SEAT worker, must lose your job or work more flexibly so that I, the shareholder, can see my investments grow. That taught us Lesson #3: *privatization is good for everyone, even if it destroys your world.*

The fourth task was to impose neoliberal logic onto the rest of the world. During the state capitalist era, the IMF, World Bank and GATT (the predecessor to the World Trade Organization) had played a back-seat role. But now the IMF kicked into life, subjecting most of Latin America, most of Africa and large parts of Asia to privatization programmes in return for debt bailouts.[8]

Mexico was the guinea pig. In August 1982 it threatened to default on $80 billion-worth of debts. The IMF's own history of the episode admits: 'The system was now at risk. The major US and Japanese banks were threatened for the first time and the European banks faced major new risks.'[9] In return for a \$4 billion bailout, Mexico was forced to impose its own version of Thatcherism: hike interest rates, cut public spending and begin a privatization programme that would sell or close 80 per cent of all state-owned factories.[10] By 1986 unemployment hit 15 per cent. The foreign debt mushroomed to \$100 billion. Real wage levels fell by at least 40 per cent in just three years.[11]

Mexico, which had staged a series of struggles for economic independence from the USA in the twentieth century, was now once again an economic colony of Washington, with a string of cheap labour factories serving the US market clustered along the border. Through Mexico, the IMF taught us Lesson #4: *economic sovereignty is impossible.*

The final challenge was to cement neoliberalism's control in Europe. In 1985, Margaret Thatcher – who had always blocked further integration in the European Community – switched track. Europe, she said, could have its parliament, its partially pooled sovereignty and its flag, on condition that it wrote neoliberalism into its key treaty, the Single European Act of 1986.

France, Mitterrand wrote, had been 'divided between two ambitions: that of the construction of Europe and that of social justice'.[12] Thatcher succeeded in imposing that choice on the entire continent. From the mid-1980s the European Community, for all its theoretical commitments to welfare provision and full employment, was practically committed to neoliberalism. At its heart stood not only Thatcherite Britain but a Germany whose elite had long ago swallowed the idea of 'as much market and as little state as possible'. Lesson #5 was that *even countries committed to the welfare state would have to deliver it using neoliberal methods.* If you want a social-market economy, you must accept privatization, outsourcing and enforced competition, and turn a blind eye to the tax-dodging of large corporations.

In fewer than ten years, the neoliberal project had reshaped the world economy. But its true achievement lay in the changes it made to the way human beings think and behave.

'The community was poor,' writes urban sociologist Janice Perlman, 'but people mobilized to demand improved urban services, worked hard, had fun, and had hope. They watched out for each other, and daily life had a calm, convivial rhythm.' That was her description of a Brazilian *favela* in the 1960s – but it could just as easily describe most working-class communities in the world back then.

When Perlman returned to Rio de Janeiro in 1999 to document the aftermath of the neoliberal transformation, things were different: 'Where there had been hope, now there were fear and uncertainty. People were afraid of getting killed in the cross fire during a drug war between competing gangs . . . They felt more marginalized than ever.'[13]

From the late 1970s onwards neoliberalism reinvented the urban slum, and forced a billion people – one in seven on the planet – to make their homes there.[14] Collapsing agricultural prices accelerated the move from the land into the cities. The state's near-bankruptcy meant there was no one to stop the new arrivals building shacks among the waterways and rubbish dumps. Slum-clearance programmes broke down: they were designed on the assumption that slums were a remnant of the past; now they would be the future.

Perlman's account of what happened in a *favela* called Nova Brasília tells the story in depth. After 1985 the major factories close down, unemployment rises massively, policing evaporates and the drug gangs move in: by the early 1990s they not only control the streets but the residents' association, after executing its last uncorrupted leader.[15] After this, she observes, the gangs effectively become the state.

What kind of person prospers in a community destroyed by drugs, violence, poverty, unemployment and insecurity? The answer is: people who can adapt to its dog-eat-dog dynamics. Those who can accept constant insecurity not as an aberration but as normality; people who are prepared to 'live for the present' and above all are prepared to look after themselves, forget community obligations, tolerate lawlessness and participate in it.

Such people were rare in the era of state capitalism, even in a poor country like Brazil. But neoliberalism created a new social archetype: the rootless, self-centred individual, focused not on the collective struggle or community activism but on the personal struggle for survival. A drug runner in the Rio *favela* might be destined to die before the age of thirty – but they could earn in a week what it would take months to earn in a factory on the minimum wage. Once you had bought your gun, looked after your family and paid for sex, what else was there to spend the money on but branded sports shoes and cheap jewellery?

As the old industries collapsed, this lifestyle, pioneered in the slums of the global south, quickly spread among young people in the devastated cities of the developed world. Rap music transmitted the new ideals of gangs, drugs and sexual violence into poor, black and Hispanic communities in the USA, but this 'bling' culture developed across many different countries and music types, becoming a kind of international neoliberal style. Ten years into the Thatcher era it was no surprise to find 'gangsta' morals, values and behaviours appearing among the disaffected young people of my home town.

By the late 1980s you have two kinds of subjectivity: a group of embittered survivors from the old system living alongside enthusiastic early adopters of selfishness, individualism and conformity. But in a world of chaos and poverty, the memory of the good times under state capitalism is strong, so the prevailing mood in working-class communities is depression and insecurity. The big negative lessons have been learned, the defeat of organized labour has been accepted, but there's still no strong, positive, universal 'common sense' for people to buy into.

For that, neoliberalism had to start functioning automatically, without major social conflict, and begin improving people's lives.

For the drug dealer in a Rio *favela* the critical ingredient to getting rich was cocaine – a globally tradable commodity whose price stays high even in a recession. For the rest of the world, the drug of choice was not cocaine but credit. And to get it flowing, neoliberalism had to become truly global.

On 4 June 1989 the Polish opposition party Solidarność won the first free elections in the Eastern Bloc. On the same day, the Chinese Communist Party (CCP) sent tanks into Beijing's Tiananmen Square to kill thousands of democracy protesters. These two events signal the beginning of neoliberalism's second phase: the globalization of the world economy, the marketization of former communist countries and the adoption of the *favela* mindset by hundreds of millions of people worldwide.

By November 1989 the Berlin Wall was down; in December 1991 the Soviet Union collapsed. While the economies of Russia, China and Eastern Europe made up only 15 per cent of global output,[16] their entry into the global marketplace would help double the size of the world's workforce: from 1.5 billion to 3 billion in the space of fifteen years. Put crudely, the same amount of capital would now have twice the amount of labour to exploit, and labour's global bargaining power would slump dramatically.

The economist Richard Freeman, who labelled this the 'Great Doubling', warned that if the USA did not adjust to the weakened social power of its own working class 'the next several decades will exacerbate economic divisions in the US and risk turning much of the country against globalization'.[17] That is exactly what happened – but at the start of the process nobody cared. Because the most important impact of globalization was ideological, not economic. The collapse of Soviet communism and the marketization of China had buried the project of the twentieth-century left for ever.

In January 1992, three weeks after the dissolution of the USSR, I arrived in Moscow to help try to revive an anti-Stalinist labour movement, working with left-wing dissidents who'd emerged during the Gorbachev era. It was a kamikaze mission. The economy was close to total collapse. Prices had risen 245 per cent in a single month; inflation would hit 2,500 per cent over the next year.[18] People lined the streets amid deep snow, trying to sell their last belongings: a single boot, a pan, their army uniform. In every hotel foyer there were women offering to sell themselves.

We held a seminar at the politics department of a Moscow university. Its professors had fled, leaving busts of Lenin, reams of statistics and the works of various Soviet leaders – which our Russian anarchist friends gleefully looted or destroyed. Unfortunately, at this very moment, the forces that would rule post-Soviet Russia were engaged in another form of looting, one infinitely more destructive: a smash and grab raid on the resources of a superpower.

A class of self-made millionaires had emerged during the late 1980s, as international trade opened up – often in the personal computer business or in the oil industry. Now an elite group among them became Russia's 'oligarchs'. They rigged privatizations to award themselves industrial plants at laughably low prices. They cornered the export markets for oil, gas and raw materials – buying at Russian prices and selling at world prices. In some cases they seized control of a company 'merely by reproducing corporate ownership documents on a home printer and then registering them with the state'.[19]

Recorded crimes in Russia rose by 50 per cent in two years. In 1990 Russian police recorded 2,800 unidentified corpses; in 1993 they found 18,000.[20] The quiet courtyards of Moscow became places you did not want to go. I saw simple automobile accidents escalate into knife fights or acid attacks. For some of the time I lived in a half-deserted student hostel, whose door my friends had simply kicked in.

All around me I could see the creation of the new human type. In the *favelas* of Brazil, it had been triggered by the arrival of cocaine. In Russia it was simply the arrival of money. Under the Soviet system, money barely functioned. You got things done through informal networks: your workplace, family, neighbours and friends. These 'kitchen table' networks were vital for social support, economic survival and the reinforcement of moral values. The sudden injection of money tore them apart.[21]

The writer Victor Pelevin captured the experience of millions of people during this forced march towards selfishness. The hero of his novel *Generation P* is forced to confront a transition 'from eternity to the present'. A system that was supposed to last for ever, and never change, had morphed into a random existence whose rules were continually in flux. He becomes a copywriter in an advertising agency, where his mentor explains the rules of neoliberalism. You borrow money, buy a Jeep, a fax machine and a crate of vodka. When your business goes bust, either the mafia kills you or they offload the loan onto a state-owned bank. Halfway through the process, 'a highly specific chemical reaction occurs inside the head of the guy who created the whole mess. He develops this totally boundless megalomania and orders himself an advertising clip. He insists this clip has to blow away all the other cretins' clips.'[22]

In that single passage Pelevin captured how neoliberalism was supposed to work for ordinary people during the transition: crooks accumulate money through crime or ripping-off the state, they recycle it via the credit system, and that generates legitimate businesses such as advertising. In the process, a new kind of person is born, attuned to the survival of the fittest. As in the *favela*, their key attribute is their willingess to tolerate criminality, thrive amid chaos and exploit the opportunities that bubble up as normal society collapses.

Sociologists labelled this new kind of person the 'neoliberal subject', because the word 'subject' in philosophy denotes the thinking human being (the 'object' being is the outside world). We could just as easily call this type of person the 'neoliberal self'.

The French sociologist Michel Foucault, writing at the dawn of the neoliberal system, understood what we would have to become in a privatized, highly competitive and impoverished society: 'entrepreneurs of the self'. By privatizing not just industries but all the risks formerly dealt with by society – vaccinations, ill health, unemployment or workplace injury – the new system forced everybody to start calculating risks at the front of their minds, in a way my parents' generation never had to.

When you're forced to keep anything at the front of your mind for a sustained period you become an expert at it. My father's generation were expert at maintaining collaborative social relationships and observing traditions and hierarchies; the neoliberal self makes you an expert at disrupting them.

For the neoliberal self, consumption – of anything, at any time, for its own sake – became a form of self-validating activity. The hero of *Generation P*, when under stress, either takes cocaine or buys something at random he can't afford: the psychological impact is the same. But as an act of communication, buying stuff is effective only if everybody else can understand the value of what you've bought: that's why the global fashion, alcohol and makeup brands became essential. Before neoliberalism, to be fashionable meant wearing clothes different from everybody else's. Now it means wearing clothes whose precise value can be understood by everyone, if necessary by sporting the word Moschino six inches high across your chest.

The neoliberal subject has, in short, exchanged security for autonomy and adopted individualism as the solution to the failure of collective action. This happened *before* mass access to the internet, let alone smartphones and 4G. By the late 1990s there was plenty of academic sociology confirming the existence of this new attitude, and showing how management theory had become one of its main transmitters in the workplace. Paris-based sociologists Luc Boltanski and Eve Chiapello documented the rise of a whole new ideology contained in the injunction to work flexibly, with 'flat hierarchies', to targets instead of the clock.[23] At the forefront, observed sociologist Richard Sennett, were the tech companies: he showed how the new, archetypal workers of neoliberalism were being formed in the loose, networked, informal and anti-hierarchical culture of the software and creative sectors.[24]

But the emergence of systematic selfishness, risk-calculation and conformist consumption tells only half the story. The neoliberal self had one more lesson to learn: that borrowing is good, and that no matter how badly financial markets crash, nothing bad ever happens.

Imagine the world economy as a poker game whose stakes are represented by global output and in which the debts of the players represent the combined debts of every country, household and company in the world.

Using this analogy, in 1991 the debts of the players match the stakes. However, by 2008 the stakes have doubled while the debts are six times higher than they were at the start.[25] Something's wrong. It doesn't matter which player is borrowing most because the entire table has a problem: most players are gambling with money that is not theirs.[26] Are you worried yet?

In the meantime, a crowd have gathered around the table and are placing side bets on the outcome of the poker game. This is the derivatives market, which barely existed in 1991 but was huge by 2008. In our poker game analogy, by 2008 these side-betters have staked *ten times* the amount of cash on the table, and much of their betting is being done with borrowed money too.[27] Still not worried?

Now let's think of the casino – in real life, the banking and finance industry. In 1991 it was raking off four cents for every dollar that passed across the table – but by 2008 it is raking off 46 cents on every dollar.[28] Somebody is underestimating the risks. If you're still not worried, your assumption must be that, if things go wrong, the casino owner will be able to issue new chips indefinitely, to cover all losses and all debts.

As neoliberalism took off, it injected financial risk into the global economy without anybody understanding the dangers. The economics profession told us that the surge of new money, loans and speculative contracts compared to economic growth was evidence of increasing perfection rather than a danger signal. While shrinking the state with religious fervour, the private sector created a mountain of unpayable debts that would have to be underwritten by the state itself.

This was an assumption that hundreds of millions of people quickly came to share: that gambling in a casino using borrowed stakes is safe, because every time it threatens to go wrong the casino issues more chips. This lesson was reinforced through recurrent cycles of financial boom and bust.

It started in Japan, where land prices tripled in the five years before 1990 and the value of shares traded on the Nikkei quadrupled. The collapse, starting in 1990, wiped 80 per cent off the stock market, stalled Japan's economic growth for the next two decades and left real wages stagnant for the same period.

But is Japan, today, a wasteland? No. The state bailed out the banks, shareholders took losses, house prices and wages stagnated for thirty years – but nobody cares. That's because, at the crucial moment, the casino did exactly what the players had assumed it would: it created new chips to keep the game going, by both borrowing and printing money. The Japanese crash taught the world a subliminal lesson: an entire country can max out its credit card, go bust and see its economy stagnate, but nothing bad ever happens.

The Asian financial crisis of 1997 came next. Thailand's currency collapsed against the dollar in July that year and, in the ensuing panic, foreign investors pulled money from Indonesia, South Korea, Singapore, Malaysia and Taiwan. Financial collapse triggered deep recessions in the real economies of the former Asian 'tiger' economies: Indonesia's output shrank by 14 per cent in a year; Thailand's by 10 per cent. But again, there was no global slump.

Next it was Russia's turn to implode. In 1997 Russia was opened to foreign finance, which flooded into the stock market and into one-year loans to the Russian government. The problem was, even as the borrowing ballooned out of control, economic growth was evaporating. The World Bank reported plaintively: 'the amounts lent grew as the fundamentals worsened'.* In August 1998 the rouble lost two-thirds of its value; the stock market nine-tenths of its value and the banking system went bust. The Russian state defaulted on the debts held by its own population; most of those with bank savings lost money. The economy shrank by 5 per cent in a year – a rare achievement even by the standards of neoliberal shock therapy.

The Russian crisis triggered the collapse of America's most successful hedge fund, Long Term Capital Management. LTCM had borrowed $125 billion against actual funds of $5 billion, in order to speculate on tiny anomalies across the global financial system. While LTCM had been borrowing twenty times its own worth from Wall Street banks, it had derivative positions 200 times its own worth, totalling 5 per cent of all such derivatives in the world. In the casino analogy, LTCM was the taker for the side bets: if it went bust, everybody else went bust. The US Federal Reserve stepped in to underwrite a $4 billion bailout.[29]

Armed with the conviction that nothing bad ever happens in a financial crash, Western investors got sucked into the dot-com bubble (1999–2001). And when that burst, after only a short pause hot money surged into the newly created market for mortgage risk, fuelling the securitization bubble that was to crash the world in 2008.

Before exploring this manic phase of neoliberalism – an era of hype-fuelled billionaires and bizarre con-artists – we should refocus the story for a moment on ourselves. What had finance done to the basic attributes of being human during the first twenty years of the neoliberal era?

Back in 1996 the title of Mark Ravenhill's debut play was too disturbing to put on billboards. *Shopping and Fucking* outraged a lot of people. It depicted the morally empty lives of young people in a consumer-driven economy in which random sex is the only solace and often reduced to a commodity itself. Some people assumed the play was a straightforward celebration of current reality. But Ravenhill's masterpiece was a dramatization of neoliberalism's deepest flaw.

If you live by market values alone you lose part of your humanity. You become self-obsessed, not just in the consumerist way prescribed

* World Bank Evaluation Brief 6, December 2008

by right-wing economists, but at a much more psychological level. Ravenhill's characters are continually engaged with the 'design' of their personalities, constantly reaching for brands and pop culture as ways to express what they suppose is unique about themselves. None of them really works: most are reliant on credit, which by 1996 was readily available even to the precariously employed.

The most shocking thing for those who understood the play was how totally this new way of thinking had obliterated any aspiration for positive change. As one character says:

> A long time ago there were big stories. Stories so big you could live your whole life in them. The Powerful Hands of the Gods and Fate. The Journey to Enlightenment. The March of Socialism. But they all died or the world grew up or grew senile or forgot them, so now we're all making up our own stories. Little stories.[30]

Many people assume that belief in socialism collapsed during the 1990s because the fall of the Berlin Wall revealed the horror and unsustainability of a state-led economy. In fact, among those resisting neoliberalism, outright supporters of the old Stalinist regimes were few. The main reason the old left's narrative collapsed was that the globalization of finance made even moderate socialism impossible – as the case of Mitterrand underscored. And at a deeper level, people came to understand that the new dynamics of capitalism had rendered ineffective the kind of workplace resistance practised by my father's generation.

In the old system you worked, you received wages, your employer made a profit and paid taxes to support the welfare state. Banks were for saving, not borrowing. If you wanted to protest, stopping production was your ultimate weapon. Even if you merely 'worked to rule' you could win a pay rise, because it was only the intricate knowledge of factory workers that kept pre-digital production processes going.

But now banks were exploiting workers directly – and successfully – by lending to them. The average net profit rate among British firms in the year Ravenhill's play came out was 13 per cent. By contrast, average credit card interest rates were around 15 per cent while 'store cards' were charging those in arrears between 18 and 30 per cent.[31] As more and more of their wages went to pay off short-term debts, people began to save less. In 1991 UK households were saving 13 per cent of their income; by 1999, this figure had fallen to 5 per cent. And unlike with a factory, you cannot go on strike against a credit card.

When I interviewed people from my home town, and asked them what was the biggest thing that had changed working-class attitudes in the past thirty years, the answer was unanimously: credit. Credit destroyed people's attachment to the one thing that had kept communities like this together for 200 years: work.

From around the mid-1990s, in a poor community, work was something you did to keep your credit card going, pay your mortgage and maintain your mobile phone topped up – it had no intrinsic worth. Under all previous forms of capitalism, for a poor person to borrow vast amounts of money was seen as stupid. Under neoliberalism in its heyday, not to borrow vast amounts of money was seen as stupid.

'Financialisation,' wrote the economist Costas Lapavitsas, 'allowed the ethics, morality and mindset of finance to penetrate social and individual life.'[32] One character in Ravenhill's play puts it even better: 'Money is civilisation . . . We haven't reached perfection. But it's the closest we've come to meaning. Civilisation is money. Money is civilisation.'[33]

The information revolution was real. By 1995 anybody with a brain could see there would be a lot of new money made and dominant market positions carved out by the new technology companies. But nobody knew which ones.

The initial public offering of Netscape's shares in 1995 saw a company with no profits launch at $28 per share, rising to $58 by the end of the first day. By the end of the year its combined shares were worth $175 billon. The Nasdaq stock exchange, which stood at 1,600 when Netscape launched, would hit 6,722 in early 2000. For anybody who bought and sold at the right time this was free money. For anyone who did the opposite, as the Nasdaq plunged during 2000–2001, it was a thrilling way to lose your savings. Except that many investors did not have savings: they were investing money they had borrowed on a credit card.

Where did the money come from to fuel this speculation? From the state. A month before Netscape launched, the US Federal Reserve cut interest rates and would go on cutting them as the Nasdaq bubble inflated. The central bank not only pumped money into the system, it provided the rationale for all the irrational decisions individuals would then make.

The Fed's boss, Alan Greenspan, assured investors that the world's stock markets – despite their soaring value – were, in fact, too low. People borrowing to invest in companies with no profits were acting more rationally than the Fed itself. 'The stock market is basically telling us that there has indeed been an acceleration of productivity,' he said.[34]

When the market crashed in March 2000, and many dot-com firms went bust, the reputation of the neoliberal economic model was in shreds. Ten Wall Street investment banks paid $1.4 billion in fines for issuing fraudulent research designed to dupe their own customers into funding companies with no profits.[35] By then Enron had collapsed, along with Worldcom, Tyco, Parmalat, Vivendi and other big corporate names, including the UK's blue-chip pension fund Equitable Life and the accounting firm Arthur Andersen. Accountancy firms had colluded in the misreporting of companies' true profitability. Banks and company executives had exploited their investors just as ruthlessly as the characters in *Shopping and Fucking* exploit each other. The legitimacy of the whole system was in doubt.

Something had to restore that legitimacy. This was the context of the global boom and bust cycle that Greenspan and other regulators stoked between 2002 and 2008. They would slash interest rates, flood the financial markets with cheap money and deregulate the banks. The 'too big to fail' logic would now be applied to the whole of capitalism itself.

But as they stoked the boom, the US elite also put the icing on the cake of the neoliberal ideology, fusing it with delusions of absolute geopolitical power. To understand why the neoliberal self has now cracked up, we must understand it became intertwined with the elite's belief in its own permanence.

4

Telegrams and Anger

If you're ever in Berlin, stand at the eastern end of the Unter den Linden thoroughfare and take a panorama shot with your phone. You will then possess a visual reminder of how badly wrong our theories about history can get.

You are surrounded by white stone columns and marble statues. The front of the opera house, the wall of an old barracks, the door of the cathedral and the entire façade of the university – all are copies of the Parthenon in Athens. The subtext is not hard to read: the Prussian aristocracy, who built this stuff 250 years ago, thought they were reconstructing ancient Greece, only bigger and better. But what did that mean to them? On 18 October 1818, the philosopher Georg Hegel strolled between these white columns and mounted a lectern to explain: it meant the end of history.

History, said Hegel, was a journey from slavery to freedom controlled by a 'world spirit', which guides humanity towards perfection in clear stages. Eastern religions had discovered the idea of the human being separate from nature. Ancient Athens had achieved something close to perfection – 'a free and unruffled ethical life' – but only for its free citizens, not its slaves. To achieve the perfect society, Hegel asserted, everybody has to be free.[1]

Now came the twist. Hegel had initially been a big admirer of the French Revolution. So big that in 1806, on the day Napoleon occupied the city of Jena, where Hegel was teaching, the philosopher hailed him as the living embodiment of the world spirit: 'It is indeed a wonderful sensation,' he wrote, 'to see such an individual, who, concentrated here at a single point, astride a horse, reaches out over the world and masters it.'[2]

But by 1818 Napoleon was defeated; all the European republics inspired by the French Revolution had been smashed; press freedom was non-existent. After thirty years of revolutionary struggle, most people who believed in freedom thought it was something you achieved by *resisting* the autocratic state. But not Hegel. You can only be free, he said, as the obedient subject of an all-powerful, enlightened state.

'The march of God in the world, that is what the state is,'[3] he declared. And with the enlightened monarchy of Prussia in 1818, the march had reached its destination. 'The history of the world travels from East to West, for Europe is absolutely the end of history, Asia the beginning,' Hegel told his students. Amid the marble columns of Berlin, under an enlightened monarch, the pinnacle of human achievement had been reached.[4]

As we know, things turned out differently. The Prussian aristocracy became as hated as the global elite are today. Their absolute power lasted just thirty years. By 1848 some of Hegel's star pupils were to be found building barricades next to these very same marble columns, in a bid to achieve democracy.

How did it all go wrong? The short answer is: a mismatch between politics and economics. At the Congress of Vienna (1815), which redrew the map of Europe after Napoleon's defeat, Prussia was designated the linchpin of a new geopolitical order. It was given large chunks of Poland and western Germany to rule, on the promise to act 'upon the most liberal principles'.[5]

But the Prussian aristocracy missed one crucial detail. While they were busy building marble copies of the Parthenon, in Britain, France and America men with money were building factories. To run factories you need people for whom obeying an aristocratic state does not seem like freedom: the liberal bourgeoisie and the working class. During the next century, conflicts between workers, capitalists and aristocrats – and between real nations and the artificial states created in 1815 – would tear apart the order created at Vienna.

Whatever else we learn from Hegel, we should have learned that declaring the end of history is usually a mistake. But after 1989 the neoliberal elite made exactly the same mistake. Francis Fukuyama's essay 'The End of History?' has been the subject of much *schadenfreude*, some of it unjust – but it is worth revisiting. Fukuyama, a state department official under Reagan and Bush, wrote in 1989:

> What we may be witnessing is not just the end of the Cold War, or the passing of a particular period of post-war history, but the end of history as such: that is, the end point of mankind's ideological evolution and the universalization of Western liberal democracy as the final form of human government.[6]

Citing Hegel as his inspiration, Fukuyama insisted that the combination of liberal democracy with free markets is an ideal that cannot be improved on. All the alternatives have been discredited – plus there are no problems in human life that markets plus democracy can't solve.

Fascism and communism were dead; nationalism and religious fundamentalism were dying. Fukuyama did not say free-market capitalism had *caused* the triumph of liberal democracy. But the arrival of the two together looked quite like Hegel's 'world-mind' at work: '[The] state of consciousness that permits the growth of liberalism seems to stabilise in the way one would expect at the end of history if it is underwritten by the abundance of a modern free market economy.'[7]

As it travelled from the pages of the magazine to the bar stools and radio talk shows of the world, the 'end of history' idea got simplified into the proposition that free-market capitalism represents a natural and perfect state, beyond which further progress is unlikely.

In Hegel's time, the elite's mistake was to design a geopolitical order but to assume the economic order would design itself. The neoliberals made the reverse error: they designed an economic system but refused to design a geopolitical system to contain it. Their ideology told them in effect that economics would shape and regulate the global order by itself.

Today the global order they established is in ruins. But because millions upon millions of people bought the 'end of history' idea, the psychological impact of it being wrong is immense. Together with self-gratification and the illusion that 'nothing bad can ever happen', belief in the permanence of neoliberalism and globalization formed the third pillar of the ideology that sustained it.

Fukuyama had warned that the end of history might be boring. It would be, he predicted, a very sad time ... the worldwide ideological struggle that called forth daring, courage, imagination and idealism will be replaced by economic calculation, the endless solving of technical problems, environmental concerns and the satisfaction of sophisticated consumer demands.[8]

And for millions of people this came true. We turned ourselves into 'human capital', calculated our financial worth, constructed our identities through mixing and matching global brands, sculpted our bodies in the gym and our faces in the beauty salon; we improved our brains with Sudoku or meditation. Gradually, the heroes and heroines in the movies we watched became one-dimensional, emotionless and bland. 'Daring, courage, imagination and idealism' are qualities we today expect to see in Hollywood villains. This bland and boring future promised by Fukuyama was only tolerable because we had achieved prosperity. Now prosperity is gone, and with it the global order, the core assumptions of the neoliberal self are undermined.

As a result of the Soviet Union's collapse, the early 1990s unfolded amid justified American hubris. There was a New World Order and it was – as journalist Charles Krauthammer memorably recorded – 'unipolar'. The US had become an unrivalled superpower, enjoying an excess of military, diplomatic and cultural might, compared to its closest rivals, bigger than at any time in history.[9]

Sure there was still chaos at the edges: in former Yugoslavia, where ethnic civil wars raged from 1991 to 1999; in Rwanda, where the first post-war genocide killed 800,000; and in Afghanistan, where the civil war dragged on until 1996. But this seemed like 'legacy chaos'. Nationalism and religious fundamentalism looked as though they were simply settling old scores, and did not seem to be forces that could shape the future. It all looked remarkably like what Fukuyama had predicted.

But by the second half of the 1990s, that is after a very short period of historical time, the consequences of trying to operate a global system without any formal political framework had become apparent. The Taliban took Kabul in 1996, installing not a liberal democracy but a religious despotism. In 1997, as the Asian financial crisis shook the global system, the Malaysian and Thai governments staged the first serious defiance of IMF economic policy. In 1998, after Russia's banking system collapsed, the Russian security elite decided to pull the plug on the liberal oligarchs, triggering the power struggle that would make Vladimir Putin president.

And across the developed world a mass, symbolic rejection of neoliberal economics had begun. The anarchists and environmentalists who'd harassed road builders and oil companies in the 1990s began leading protesters in much larger numbers against – and sometimes over – the fence lines of global economic summits. In Seattle in 1999, Prague in 2000 and culminating in the bloody riot of Genoa in April 2001, a new, radical anticapitalist movement shook the confidence of the neoliberal elite.

In 2001, three events occurred that should have killed the 'end of history' delusion stone dead. First, the terrorist attacks of 9/11 – organized by the very forces that America had allowed to control Afghanistan; then the Enron bankruptcy, which revealed systemic misregulation and corruption across corporate America; finally, amid mass unrest, the bankruptcy of Argentina in December.

Looking back on that year, whose events I reported for the BBC, it is obvious why the chaos at the edges of the system was treated as background noise. Neoliberalism looked like it could survive the revival of geopolitical chaos. Provided all the wars and terror threats now rising remained at the periphery, the illusion could be maintained: history is over and nothing bad can ever happen, so long as we crack down hard on the sources of chaos.

In response to the traumas of 2001, however, a change took place in the thinking of the US conservative elite, at the level of both economics and geopolitics. Federal Reserve boss Alan Greenspan abandoned any attempt to rein in the financial markets, signalling instead the permanent availability of cheap money that would fuel the mortgage bubble of 2003–8. For Greenspan, the economy's bounceback after 9/11 also proved a mental watershed, confirming his view that information technology had created a world impervious to financial danger: 'After 9/11 I knew . . . that we are living in a new world – the world of a global capitalist economy that is vastly more flexible, resilient, open, self-correcting and fast changing than it was even a quarter of a century ago.'[10]

In the Pentagon, key Reagan-era hawks were now in control of US foreign policy. Authorized to wage a rules-free 'war on terror', they expanded it to pursue two long-cherished goals: to normalize torture and detention without trial under the US Constitution; and to invade Iraq, in order to seize its oil and 'stabilize' the Middle East. They would make their case by lying through their teeth – but this was not mere mendacity.

At an intellectual level, something subtle but decisive happened to the thinking of figures such as Alan Greenspan, Donald Rumsfeld and their milieu after 9/11. At the very moment when the frailty of the global order was revealed – in the markets, in corporate governance and in geopolitics – they decided to take crazy strategic risks, justified by the delusion that they could shape all reality to their willpower.

The delusion was outlined in the chilling words of Bush aide Karl Rove to a journalist in 2002:

> We're an empire now, and when we act, we create our own reality. And while you're studying that reality . . . we'll act again, creating other new realities, which you can study too, and that's how things will sort out. We're history's actors . . . and you, all of you, will be left to just study what we do.[11]

This was a theory of absolute supremacy, of the unlimited capacity to act.

The delusion was communicated to millions of people, above all through the best-selling works of *New York Times* writer Thomas Friedman. Friedman revealed the true logic behind the attack on Iraq: 'America needed to hit someone in the Arab-Muslim world,' he wrote. 'Smashing Saudi Arabia or Syria would have been fine. But we hit Saddam for one simple reason: because we could.'[12] Iraq was a demonstration that the USA could take arbitrary action, killing hundreds of thousands of people with impunity.

Anyone who has studied logic should by now be able to spot the problem with neoliberalism after 2001. Its elite was conflating – often deliberately, as Rove's comment indicated – facts and a wishlist; *dressing up a claim about 'what ought to be' as a claim about 'what is'.*

Until 2001, the emergence of globalized trade and financial markets was greeted as an unstoppable, one-way historical process. One of the most impolite things you could do in the company of neo-liberals was to suggest globalization was merely a policy, and could be reversed. But after 2001 the world no longer fitted neoliberal theory. So it had to be forced to.

This was the source of a new coerciveness that entered elite thinking around the year 2001. It is what links the geopolitical recklessness of Bush and Blair over Iraq to the Fed's cheap money frenzy between 2002 and 2008. Both were attempts to sustain the 'end of history' illusion. Both relied on systematic lying. Both led to disaster.

The Iraq War set off a chain reaction that would destroy the unipolar order. Iraq taught both Moscow and Beijing that any unipolar system would be run in an overtly neo-colonialist way, with the 'hidden fist' of the F15 bomber no longer promoting the hidden hand of market forces in the abstract, but primarily furthering the interests of US corporate monopolies like Halliburton and ExxonMobil. In a century when natural resources will become scarcer, it taught the world that America was prepared to go to war in order to secure them. It taught authoritarians everywhere that Western 'humanitarian intervention' was a sham. And it empowered those within both China and Russia arguing for conventional rearmament: each more than doubled their defence spending in the following decade.[13]

All the main players in geopolitics, including the EU and Japan, realized that America's unipolar power was waning. In that situation, it was logical for everyone to think about an active geopolitical strategy of their own, to fill the void left by the USA. And Vladimir Putin was not just thinking about it: he was obsessed with it.

History, instead of ending, was about to accelerate. But Fukuyama's generation was to face a much bigger problem than Hegel's. The geopolitical order designed in 1815 got torn apart by economic growth, technological innovation and the expansion of democracy: in short, by progress. The global economic order born in 1989 is being torn apart by economic stagnation and by forces opposed to science and democracy. That is the opposite of progress.

The most startling thing about the pre-2008 financial frenzy is, with hindsight, how ideologically driven it was – reflecting neoliberalism's switch from 'this is how things are' to 'this is how things ought to be'.

The rationale was, said politicians, that poor people *ought* to have access to mortgages on worthless property in order to 'financially include' them in the system. With the psychology of young people attuned to brands and consumption, the supply of credit *ought* to keep going, no matter how badly wages stagnated. House prices *ought* to rise so that the baby boomers could pass on their wealth to the impoverished millennials. Stock markets *ought* to flourish so that the now retired hippy generation could afford organic food and holidays in Costa Rica. Having forced into people's brains, through very painful lessons, the belief that free-market capitalism works, it had to be *made* to work by pumping it full of money.

The result was the rise of a massive global imbalance between countries that export and lend, and those which import and borrow. By 2006 the size of this global imbalance, as measured by all the current account mismatches, was 5.5 per cent of world GDP.[14] Most economists said: don't worry – these are just the growing pains of globalization. But as money flooded through the global pipework of the financial system with nobody to regulate it, the system was bound to develop blockages and leakages. There was nobody to control the flow. French economists Anton Brender and Florence Pisani summed up succinctly how the absence of global regulation made the 2008 collapse inevitable: 'The consequence was terrible: the only force that could finally rein in the continuous deepening of the global imbalances was the collapse of globalised finance.'[15]

It suited everybody for there to be a finance system without governance, and for the global order to be based on the arbitrary projection of American power. Both arrangements became inseparable from the neoliberal project and both helped destroy it.

The warnings were there. But the elite ignored them. For by now the tenets of neoliberalism had morphed into a theory of human nature.

The term 'economic man' was first suggested by John Stuart Mill as a thought experiment. In order to search for consistent patterns in our economic behaviour, Mill and economists who followed him imagined human beings stripped of all other attributes. *Homo economicus* was defined as a selfish individual pursuing his own interest, seeking maximum benefits for himself and operating with perfect knowledge. But nobody actually claimed such two-dimensional individuals existed: economic man was an abstraction.

Mill and his generation subscribed to the liberal view of human nature: they believed that human beings are naturally competitive – but they also understood that real people are influenced by religion, ethics, the desire for luxury and leisure.

Yet neoliberalism turned the thought experiments of nineteenth-century economics into claims about reality. It did this, first, at the level of academic theory. Neoliberalism's theorists defined all aspects of human nature as essentially economic; they defined the essential feature of a market-based economy as competition, not exchange; and they redefined the worker as 'human capital'. Gary Becker, the cult hero of neoliberal academia, said that any decision taken rationally can be modelled as if it were an economic choice – whether it be crime, sex or voting. He produced a mathematical formula showing how optimal levels of crime could be achieved by making the risks outweigh the rewards.[16]

Then, over a thirty-year period, the elite used theories developed by Becker and his followers to impose economic imperatives onto real people, attacking any remaining impulses towards collaboration, solidarity and altruism. They did so using laws, management techniques, financial incentives, propaganda and outright force. At their most ambitious they even tried – and are still trying, through companies such as Uber and Airbnb – to replace corporations, states and organizations with random collections of individual 'economic men'.

But once the policy elite had unleashed the power of new routines to reshape our thinking, a change began at the micro-level of human life. It turned out that being a good economic man or woman makes you a very inefficient citizen. As Michel Foucault pointed out, while the *homo economicus* imagined by nineteenth-century liberals was a person to be left alone by the state, and left free to choose, under neoliberalism he is someone to be managed: 'someone who is eminently governable'.[17]

The creeping commercialization of culture, sex and leisure has always been inherent in the logic of capitalism. But in the neoliberal era it was no longer gradual. Something unprecedented happened to the typical person, and to the average set of ideas in people's heads during the past thirty years, reaching its high point during the pre-2008 speculative frenzy.

Ideology, as understood by critics of capitalism, is a set of ideas that masks reality. It is created by what we see and feel, and reinforced by the fact that the elite controls the flow of information. So, for example, in the Soviet Union people were told (and told each other) that they were living under 'actually existing socialism', whereas the reality was dictatorship, poverty, misery and inequality.

Ideologies, typically, are defined against clear, visible alternatives. Insofar as they mask a deeper, hidden truth, educated and inquisitive people can think their way out of them – especially if there is an organized counter-power like the labour movement, which warns: treat everything your boss says as bullshit.

What made neoliberalism different is the way it overcame this: it created a reality in which it became impossible to imagine alternatives. Educated and inquisitive individuals found it increasingly impossible to think their way beyond it.

When McDonald's first opened in my home town, it created awkward situations. In all other shops, cafés and department stores, customers chatted to those serving them: they knew them; they would ask after their families, or discuss where to meet up on Saturday night. In McDonald's there was a new routine. The server was working from a script. It was easier if they ignored the fact they'd gone to school with you. They could not easily shoot the breeze. Though a few people on both sides of the counter rebelled at first, in the long term it worked best if both you and the customer stuck to this new, impersonal, corporate routine. Your burger arrived quicker and nobody got the sack. The more you participated in the performance, the better you felt.

In this long global process, across billions of small, everyday transactions and routines, neoliberalism became something bigger than an ideology. It became what political scientist Wendy Brown calls an 'order of normative reason' – more like a religion, or an Excel spreadsheet, whose logic is unchallengeable.[18]

In the Soviet Union and Mao's China, the ruling ideologies were easy to puncture. People were told their societies were the most prosperous on earth – but all they had to do was watch a Hollywood movie, or fly to Los Angeles on a trade delegation, to understand this was a lie. That's why societies with fragile ideologies always try to limit contact with the outside reality: once you compared Soviet ideology to Western reality it shattered.

By contrast, neoliberalism was so deeply embedded that the more you compared the ideology to reality, the more it seemed correct. The feedback loops between enforced competitive behaviour, credit dependency and short-term prosperity were strong. The only condition was that you kept your emotions, ideals and any remaining ethics in a separate compartment – away from the central activity of working, trading and competing.

However, sometimes that is not possible. For neoliberalism, the equivalent of the Soviet defector's trip to LA was a visit to our 360-degree selves: to the inner, rounded, ethical and social human being our mothers gave birth to. For neoliberalism, the equivalent of the Soviet travel ban was the continuous reward for *homo economicus*-style behaviour, and punishment for behaviours that defy economic logic but promote human values. That is why the security apparatus of the Western world declared war on environmental protesters in the 1990s, and their successors in the anti-capitalist movements.

The neoliberal project was in practice an assault on humanism. It enforced the reduction of human nature to economic competition, and it suppressed all attempts to experiment with alternatives. Once its dynamism disappeared in 2008, the 'order of normative reason' collapsed. That's the explanation for why so many people, so easily, were able to revert to the logic of ethnic nationalism, misogyny and anti-science: their mental defences against these ideologies had been destroyed.

What began in 2008 is not just a global economic crisis but a crisis of the neoliberal subject. One by one the illusions built up over thirty years, around which millions had structured their lives, vanished.

The belief that complex financial systems enhance the stability of the real economy? Over. The assumption that nothing bad ever happens when a speculative bubble bursts? Dust. The idea that politics is about technocratic parties arguing with each other about minor details from here to eternity? Gone. The religion of cheap credit? Debunked. The dogma that if everybody competes with everybody else, things can only get better? Disproven in every welfare office, at every food bank, with every sorry doorway filled with a human being huddled in a sleeping bag.

But the lost illusions are only half of the problem. By reducing everything to economics, and by authorizing systemized lying of the type that killed Lehman Brothers and justified the Iraq debacle, neoliberalism absolved an entire generation from moral judgements. So long as you obeyed the performance rituals of neoliberalism – at work, the gym, the wine bar – the system was neutral as to your ethical beliefs.

Neoliberalism became a system of performance: a kind of ritualized theatre. Performative behaviour is easy to standardize and measure in market terms. Your department has met the benchmark for best practice in hiring women and minorities? Give yourself a tick. Who cares if you secretly think whites are biologically superior to blacks and men to women? The liberal assumption was that, as economic growth and technological progress made things better for everyone, the reactionary prejudices of some individuals might fade away. Even if they did not, it was of secondary importance so long as such beliefs never intruded on economically rational decision-making.

However, in societies based on performance rituals, it is possible for large numbers of people to develop transgressive ideas in defiance of those rituals, often in secret. When the rituals no longer deliver prosperity, and the given ideology no longer makes sense, people search for new ones that match their experience. Today, you only have to watch Twitter for half an hour to understand how that experience has led some individuals to racist, misogynist, anti-Semitic or Islamophobic conclusions.

When sociologists describe the neoliberal self, they often give a list of behaviours and attitudes drilled into us by the market: respect for money, the tendency to define freedom as a form of consumer choice, the willingness to view the self as 'human capital', the obsession with celebrities and brands. But it was always about more than this. The neoliberal self was intrinsically rooted in the idea of geopolitical permanence, and in the absence of economic alternatives.

To understand the acute crisis of identity that millions of people are now living through, we have to trace the process whereby both the geopolitics and the economics fell apart at once.

5

The Crack Up

Before 2008, neoliberalism's promise was: things will be like this for ever, only better. After 2008 it was: things will be like this for ever, only worse.

Threatened with a depression deeper than that of the 1930s, those in power acted to stop it. But their actions contradicted the narrative they had fed people over the preceding thirty years. That's why the narrative has fallen apart. In the space of a decade the neoliberal self, so painstakingly coerced into existence, has seen its habitat destroyed as ruthlessly as that of the Sumatran tiger.

The journey from Lehman Brothers' collapse to Brexit and Trump's victory is not primarily about economic hardship. It is about the elite's refusal to learn from failure once their story fell apart; their switch from coercion and propaganda to outright violence and manipulation against those who wanted to change the system; and the left's inability to project a clear alternative.

Though right-wing populist movements are accused of wanting to re-run the past, after 2008 the neoliberal project itself became a kind of nostalgia movement for the euphoria of the 1990s. The radical left has so far failed because it, too, did not differentiate its project sufficiently from the past. As a result, the 2008–16 phase of neoliberalism played out as a competition between three different kinds of nostalgia.

In October 2008 the British government used £500 billion of tax-payer money to save its banks. The US TARP programme pledged $700 billion; its French equivalent €360 billion. Ireland pledged amounts so large that its public finances were implicitly wrecked. All told, including insurance and guarantees, by November 2009 the USA, Britain and the Eurozone had injected around $8 trillion into their banks.[1]

The bank bailouts transferred a lot more than money. They transferred all the risks of banking and finance onto the state, while allowing investors and managers in the banking sector to go on privatizing the rewards. This blew apart the fundamental tenets of the ideology that had been crammed into people's brains: that government intervention causes more harm than good, and that the market always corrects itself.

Meanwhile, the speed at which the system collapsed undermined a core assumption of professional economics: that complexity equals safety; that spreading risks across sectors, time zones and asset classes allows the whole world to take the strain if something goes wrong.

The Bank of England's chief economist, Andy Haldane, expressed it bluntly. Mathematical models had pointed to the stabilizing effects of complexity. Instead, the system had 'shown itself to be neither self-regulating nor self-repairing. Like the rainforests, when faced with a big shock, the financial system has at times risked becoming non-renewable.'[2]

When a central banker tells you that capitalism is, like a rainforest, at risk of extinction, you should listen. Those who were listening understood the basic message: the whole system had been based on lies. And with the central lie exposed, the rest of the story disintegrated.

Since 1992 the European Union had banned state aid to private sector companies. In the USA, both Reagan and Bush had tried to 'eliminate industrial policy wherever they found it'.[3] The idea of supporting steel plants or auto factories to maintain employment, expertise and security of supply had been derided. After 2008 states would support private companies wholesale. Not only the banks and insurance companies but the auto-makers and engineers, their financial offshoots and even companies like French toy-maker Meccano were given taxpayers' money to survive.[4]

As industrial production, trade and employment slid backwards across the developed world, countries were forced to cut taxes and at the same time raise spending. This is called a 'fiscal stimulus' and came straight out of the textbook of the state-capitalist era: in a crisis, said Keynes, you borrow and spend. But most politicians and econo-mists had built their careers on telling people Keynes's textbook was wrong. In addition, the states required to hike borrowing were already carrying large debts – so their debt-to-GDP ratios would rise above limits dictated by their own economic rules and doctrines, and in the EU's case by law.

By 2010, with the immediate crisis over, these high levels of debt triggered the demand for tough austerity. From whom? The bailed-out banks – which now threatened not to lend to governments unless they attacked their own populations. As governments from Ireland to Greece were forced to cut spending on pensions, wages and services, growth again stagnated, precipitating the Eurocrisis of 2011–12. People all over the world, watching the impact on a country like Greece, could now see the human cost of neoliberalism: rising suicide rates, deteriorating standards of health and social care, stagnating growth. The UK joined in, with the Conservative–Liberal coalition cutting public spending so hard that there, too, growth flatlined.

Ultimately what kept the world economy afloat was the turn by cen-tral banks to quantitative easing (QE): cutting interest rates to zero and creating new money from nothing. By 2018 central banks had injected $20 trillion of new demand. Free-market ideology told the central bank-ers they should never spend the newly printed money on real things, such as infrastructure, healthcare or university fees. So they relied on an indirect effect to stimulate the economy. If you buy up the safest loans on earth until there is a shortage of them, you force investors to move their money to less safe containers: stocks, shares, property, commod-ities, gold and Bitcoin. The price of these assets will then rise and the profits eventually trickle through to the real economy: new malls and office blocks get built, new businesses get formed, new millionaires are made and they need to buy the latest Swiss chronometer.

But for the poor and the lower middle class there was a downside. First, QE reduced the incomes of people living on private pensions – since there was now minimal interest to be generated from government bonds, which pension funds have to hold. Then it stimulated another massive asset bubble: real estate prices, rents, stock markets and com-modity prices soared. So did the value of finite stores of value – gold, from 2009 to 2012; and Bitcoin from 2012 onwards. This was good if you were a property speculator, or a mining boss in Angola, or a Russian oligarch – but bad if you had only wages to live on, since wages were barely rising at all.

Between 2007 and 2015 real wages fell in Greece, Italy, Portugal and the UK. Japan, Spain, France and the USA achieved less than 1 per cent annual wage growth in the eight years following the credit crunch. In no developed economy did wages keep pace with GDP growth: the long slide in the 'wage share' of developed economies, compared to profits, accelerated.

Stagnating wages, combined with rotting infrastructure, deteriorating public services and the rising cost of everything from university fees to rents: this was the lived experience of the 'recovery' in the countries that had pursued neoliberal policies to the max. Set alongside the spectacular rise in the price of shares and luxury apartments, it forced people to unlearn all the lessons neoliberalism had taught them during its rise: people's everyday experience began to teach them neoliberalism worked only for the rich.

After the state bailouts, writes the British economist William Davies, neoliberalism became 'literally unjustified'. It became 'a ritual to be repeated, not a judgement to be believed'.[5] An entire generation who had been told that states should be small and inactive saw them act fast, massively and arbitrarily – with no attempt to theorize or explain.

There were only two rational conclusions: abandon the neoliberal model, or reshape it in a form whereby every state is fighting for a piece of a smaller pie – an option I've labelled 'national neoliberalism'. Trump and Brexit are clear examples of this latter option; so, too, was Germany's decision to smash Greek democracy in 2015. It all comes down to imposing the cost of crisis on someone else, so that your own variant of neoliberalism can survive.

As a result, economic nationalism has returned, but not in the state-capitalist form people expected it to assume. It has returned in the demand for the repudiation of trade treaties, weaker global institutions and cross-border regulations. To understand why this new, national form of neoliberalism is rational for parts of the elite, let us survey the long-term sources of growth, past and future.

*

In 2015 economists at the Bank of England attempted to show how, for thirty years, the world economy has been driven by a kind of growth that will not sustain itself in future. From 1980 to 2000 there were only two drivers of global growth: an increased workforce and growth at the 'frontier' of productivity, which meant credit growth and rising levels of education rather than pure technological innovation. During this period – when poor countries were being forced to obey the strictures of the IMF – the global south was actually a drag on global growth figures.

After 2000, say the Bank's economists, things changed. The take-off of industrial production and innovation in China, and the refusal by countries in Asia and Latin America to obey IMF austerity demands, pushed 'catch up growth' in the global south from negative to spectacularly positive between 2000 and 2010. Catch up growth is a process whereby Croatia becomes more like Italy, Turkey becomes more like Croatia and so on. It involves building infrastructure and raising the educational level of the workforce. But it becomes harder to do as time goes on.

In the same period – 2000 to 2010 – the contribution of an expanded workforce to growth slowed. As for productivity growth, that turned negative. Over the entire thirty-year period, say the Bank's economists, pure technological innovation drove global growth by precisely minus 0.2 per cent. That means less than not at all.

This has profound implications for the future relationship between human beings, markets and machines. For now, however, we can state the root cause of the 2008 crisis in a form short enough to tweet: technological innovation was no longer delivering growth sufficient to make all the borrowing rational.

If this is right, it means that neoliberalism was not a solution to the breakdown of the state-capitalist system designed by Keynes: it was a work-around. It relied on credit, growing population, rising education and urbanization to fuel growth. But, as the Bank of England economists' predictions show, all of these things are finite.

In the next thirty years, as population growth slows and the gap between poorer and richer countries narrows, it will be logical for countries to compete for the remaining growth in the world, still trying to make the neoliberal recipe of deregulation, asset wealth and small states work, only now with the added paprika of ethno-nationalism.

By early 2016 the central banks had managed to keep the world economy on life support for eight years. And, they assured each other at the G20 financial summit in Shanghai, they could go on doing it for a long time.[6] By printing money you can keep an economy on life support for ever. The problem is, you can't keep an ideology on life support. The human brain demands coherence.

People wanted to know when life would get better for them, not just for yacht-owners. In the austerity-ravaged countries, they wanted to know when the pain would end. The young wanted to know how they were supposed to pay off mountains of student debt on permanently stagnating wages, and how they would save for retirement now that company pension systems were closed off. The elites of the G7 countries no longer had answers.

As a result, the more people compare it to the reality of their daily lives, the more neoliberal ideology seems like a lie. Instead of free exchange, it is increasingly reliant on enforced competition: between schoolkids, universities, cities, workers, tenants, cab drivers – and the purpose of the competition is always to get the Ordinary Joe to do more for less. Instead of a free market filled with entrepreneurs, the business landscape is now dominated by monopolies on a scale unseen during the state-capitalist era: Google, Facebook, Apple, Amazon, Alibaba, Tencent and their like – always structured so that the management has more power than ordinary investors and always prepared to destroy or acquire potential competitors.

Instead of social mobility, access to the six-figure salary – via top universities and professional qualifications – has in many countries become hereditary. The 'barrow boys' who made fortunes in the London financial markets in the 1980s were replaced by the sons and daughters of millionaires. Acting, journalism and the law, vocations which once functioned as routes out of the working class for intelligent kids, have also became dominated by privately educated children of the rich.

Meanwhile, as the developed world stagnated, the elites and middle classes of the emerging markets carried on rising. From the 1980s, the elites told people that globalization was a way for rich countries to get richer by breaking into the markets of poor countries; people with darker skins would do the dirty manufacturing work; the citizens of the G7 countries would have the high-paid, high-skilled jobs. After 2008, the illusory nature of this promise was clear. The new millionaires haunting the Rolex counters of the world were from China, Russia, Kazakhstan or Angola. Global trade, which had been sold as a way of enriching the working class of the developed world, looked more and more like a way of impoverishing them.

But what finally broke people's will to tolerate neoliberalism in its global and democratic forms was something else: the technological empowerment of their emotions.

In the novel *Howards End*, written just before the First World War, E. M. Forster uses two middle-class families to illustrate contrasting approaches to life. One favours the inner life of culture, emotions and relationships; the other, a life of action, business and conflict, which Forster summed up in the phrase 'telegrams and anger'.[7]

During the neoliberal era, focusing on the inner emotional life and personal relationships was the way millions of people coped with an increasingly chaotic world. But after 1995 the technology they used to do so was radically different from any other generation. Forster's characters used the fountain pen, the paintbrush and the keyboard of a piano. We, by contrast, conduct our inner life through connected information devices – and the outbreak of systemic crisis coincided almost exactly with the point where these devices had enabled us to form mass, mobile, social networks.

For the first two years after 2008 there was remarkably little resistance. There was instead a kind of pervasive mental and verbal discontent. But it could not help itself from spreading and connecting: self-expression, even in its most introspective forms, was forced into the world of action by the technology we used. Forster, who had lectured the pre-1914 generation that they need 'only connect' their desires to their actions, would have been pleasantly surprised by the result.

From mid-2009 the resistance began. When it burst onto the squares and streets in 2011, the neoliberal elites responded not just with 'telegrams and anger' but with repression, censorship and violence.

Summer 2009 saw the protests coordinated via social media sweep Iran. A wave of student occupations hit the USA over sharp hikes in tuition fees. In October 2010 French youths, a mixture of students and the unemployed, 'smashed storefronts and threw up roadblocks across France, staging running battles with riot police'. The trigger? A decision to raise the retirement age to sixty-two.[8] A month later British students occupied their colleges and, together with kids from high school and the unemployed, stormed the headquarters of the Conservative Party and then paralysed Whitehall in three chaotic demonstrations. The detonator here was a hike in university tuition fees from £3,000 to £9,000 a year. The response in all these cases was disproportionate police violence.

Then, a protest movement against poverty and corrupt policing in Tunisia toppled its dictator, Zine El Abidine Ben Ali. On 25 January 2011 the occupation of Cairo's Tahrir Square began and the Arab world ignited: the civil wars that were to scar Yemen, Libya and Syria all started as peaceful civil protest movements within days of the Tahrir occupation. Social order in Bahrain was rocked so badly that Saudi Arabia deployed its armed forces to prevent its spread.

The unifying feature of the protests was their use of networked social media. Egypt's uprising was organized on Facebook; the mayhem generated by the London student occupations rolled live on Twitter. Long-suppressed bloggers and citizen journalists came into the open across the Arab world. And now the state-controlled Arab media and the Western broadcast channels all faced the same problem: the narrative of the protest was outside their control. Hand-held video of police atrocities, and of awe-inspiring courage by young protesters, was speeding from country to country. It was impervious to censorship and editorial self-restraint. So was the central slogan of the uprising: 'The people demand the fall of the regime.'

Some people actually thought the uprisings had been caused by Facebook. They held banners in Cairo's Tahrir Square saying 'Thank You Facebook' and painted its logo on the walls. In fact, the critical moment in Egypt was when Mubarak shut down Facebook and people took to the streets. What drove the global protest movement was a combination of three things: economic grievance, networked communications and a methodology of protest formed by theories which explicitly rejected the old, hierarchical methods of socialism, trade unions and Arab nationalism.

Then, from May to July 2011, came the occupation of the squares in Europe, drawing in millions of people in Spain and hundreds of thousands in Greece. In July, the youth of Israel – both Arabs and Jews – joined in, occupying Rothschild Boulevard in Tel Aviv in protest against poor housing and youth poverty. On 17 September 2011 the Occupy Wall Street protest took New York's Zuccotti Park, triggering the global Occupy movement, whose actions touched hundreds of cities worldwide. After the Zuccotti tent camp was smashed up in November, Russia picked up the baton.

That December, a wave of demonstrations protested against the rigging of the Duma election in favour of Putin's United Russia party. They pulled in tens of thousands of people, peaking at 120,000 on 24 December. This was a mixture of the new networked youth we'd seen on the Occupy protests, with the remnants of the oligarch-run liberalism of the 1990s and a fair smattering of nationalists and xenophobes. After more large demos in February–March 2012, the movement subsided in the face of police violence, the arrest of its grassroots leaders on fabricated tax charges, media vilification, state cyber attacks – and the fists of neofascist mobs supporting Putin.

Though each of these uprisings had a specific national character, they had all been provoked by pervasive injustices linked to neoliberalism. In Britain and peripheral Europe the issue was austerity; in the USA the government's unwillingness to take on Wall Street. In the Arab world the protests were driven by rising prices – fuelled by quantitative easing in the developed world – and the arrogance of presidential kleptocrats whose sons, like Saif Gaddafi and Gamal Mubarak, had become made men in the neoliberal mafia. In Russia, despite a decade of economic development, it was fuelled by outrage over the price society had paid: blatant kleptocracy, organized crime and a hollowed out democracy.

In each case, though the energy to act had been accumulated on the internet, so had a shared vision of a society based on equality, shaped around free, networked individuals.

*

The sociologists who coined the term 'networked individual' used it to describe changes in behaviour during the 1990s, as a combination of flat management structures, suburban living patterns and mass access to networked devices began to impact the way we live. We were, said Barry Wellman, moving away from living in groups and hierarchies towards living in networks. Manuel Castells took the concept further, arguing that the information age has produced a whole new culture, power structure and self-imagery among people exposed to it. Wellman thinks the phenomenon reversible, Castells not: he argues that, just as you cannot de-electrify a country, you cannot de-network a society.[9]

In November 2010, watching a leaderless mass of sixteen-year-olds from the deprived suburbs of London swarm into Whitehall, fight the cops and dance on top of a police car, it struck me that Castells's theory was correct. Whatever the specific national or cultural context, the behaviour from London to Cairo to Athens to New York was new and similar.

Wherever they could, the protesters occupied public space, setting up camps or temporary assemblies. On the internet they had evolved not only a political programme, or a schedule of action, but a new model of human interaction – based on consensus, diversity and the horizontal power structures networks encourage. Instead of simply demonstrating for social change and then going home, writes Castells, they created a temporary model of the society they wanted to live in, on the nearest iconic public square: 'The Internet provided the safe space where networks of outrage and hope connected. Networks formed in cyberspace extended their reach to urban space.'[10]

Castells describes this surge from cyberspace to the streets as a voluntary act. Looking back it seems to me more involuntary: as if the technology itself forced all attempts at private self-expression to veer away from introspection and out into the world. When you are passive and depressed you might share that feeling with only a few people; when you are mad as hell you want everyone to know. With social media you had the means to tell them.

Common to most of these movements was a strong critique of consumerism and the submergence of all divisions and identities – class, gender, religion – into the euphoria of the occupied space. They held long, slow, deliberative assemblies in which ideas were explored obliquely, where ideological polemics were sidelined, where aggressiveness and hierarchical behaviour were discouraged. Prevalent in these discussions was the belief among young people that they no longer had a stake in the future of the system as it was configured. Once the wave of protest had taken down Ben Ali and Mubarak, it was obvious it had at least the force of the contagious revolution that swept Europe in 1848, and that its effects would be long lasting. To suppress it, one state after another resorted to militarized policing.

On 29 June 2011, I watched Greek riot police fire more than 1,000 rounds of tear gas into the tent camp in Syntagma Square. Their aim was to disperse a vast, milling crowd of protesters, who'd assembled to resist the passage of a second austerity package on the orders of the IMF and ECB. Apart from a hard core of anarchists, it was clear that most people were not there to use violence, but to use their bodies as obstacles to the cops. In the narrow shopping streets below Syntagma I saw small-business owners close their shops, wrap their faces in damp cloths and build barricades alongside the radical youth. 'What do you do?' I randomly asked people. 'Interior designer, concert pianist, furniture salesman' came the answers. People with ruined businesses alongside graduates with ruined futures.

Militarized policing proved highly effective at destroying peaceful protests and victimizing activists. But, as the images spread across the Facebook and Instagram accounts of a generation, they revealed to millions of people the ultimate truth: neoliberalism is the logic of the market imposed by violence.

The young British journalist Laurie Penny, reporting from Greece in 2012, described this moment of realization:

> The first time you get a kicking from the police, or see your friends hurt and arrested, and you realise where the lines of power are really drawn, and nothing has changed but everything is somehow different. It's a vital part of our education, but once you've learned that lesson once you don't need it again. I smoke a cigarette I don't really want, and I am angry, I am angry, I am angry.[11]

In Europe, the USA, Tunisia and Egypt the offensive phase of the street revolts was over by 2012. But in other key countries the dynamic unfolded later: the following year Brazil, Ukraine and Turkey each saw mass movements against corrupt, autocratic governments. Meanwhile, in Syria, Libya and Yemen the revolts of civil society had turned into civil wars, creating a magnet for geopolitical actors beyond the control of any protest movement: jihadism, US strategic power and a coalition of states and movements aligned to Russian foreign policy.

These shocking events – which plastered imagery of torture, rape and indiscriminate killing of civilians across the social media – are sometimes said to have inured the global populace to violence. In fact, such reporting has made a big difference; it has made people well aware that large-scale and indiscriminate violence is the default option of kleptocratic elites – and that, when challenged, such elites will abandon their commitments to the rule of law and democracy.

Between 2009 and 2015, in country after country, resistance movements offered the world a glimpse of the progressive future. But the ruling elites, in general, said: 'no thanks'. They suppressed progressive alternatives using all the hardware and legal powers they had created during the upswing of the system. However, twenty-first-century authoritarianism could not be simply about preserving a crumbling status quo. The world was about to find out that if you don't want the future, what you're going to get is the past.

6

The Road to Kekistan

On the night of 11 August 2017 around 250 men clad in polo shirts and khaki trousers staged a Nazi-style torchlit march through the American university city of Charlottesville, Virginia. They chanted 'Blood and Soil', 'White lives matter' and 'Jews will not replace us'. Their aim was to protest against the planned removal of a statue of Confederate general Robert E. Lee, by the city council. The next day, a much bigger march, labelled Unite the Right, sparked violence across the city. The police lost control of the march, leading to both the city and the state declaring states of emergency.

The marchers included four neo-Nazi groups, the Ku Klux Klan, members of the Identitarian movement and at least three self-styled militias carrying semi-automatic weapons.[1] After the police drove antifascist protesters into a side-street, James Alex Fields, who had been carrying a shield bearing the logo of Vanguard America, one of the fascist groups, drove his truck into them, killing the protester Heather Heyer and injuring nineteen others. A police helicopter monitoring the violence crashed, killing two officers.

Two hours after the attack, Donald Trump made a statement condemning the violence 'on many sides'. Three days later, in an unscripted outburst, Trump claimed there were 'very fine people' on both sides, slammed attempts to remove Confederate monuments and attacked the 'alt-left', who he claimed had been 'very, very violent'. Former KKK supremo David Duke immediately tweeted his support for Trump. Steve Bannon is said to have called the outburst a 'defining moment', during which Trump stood with 'his people'.[2]

The Charlottesville event was just one of a rolling series of alt-right outrages which continue both in the USA and beyond it. But it contained in microcosm most of the elements we need to understand the challenge of the new fascism to the frayed democracies of the developed world.

A good place to start is with the banners waved. Most of them were what you might expect: the Confederate flag, the swastika, the black sun banner of the Nazi SS, the black cross of the Southern Nationalist movement, which wants to revive the Confederacy, and the Spartan shield logo adopted by European Identitarians as a symbol of their opposition to immigration. But one flag was new to people who had not been studying the far right closely: a black and green parody of Nazi Wehrmacht's war flag with the logo of a website called '4chan' in place of a swastika. This is the flag of 'Kekistan', a fictional right-wing Utopia dreamed up to 'troll' liberal and progressive Americans, using subverted symbols from popular culture.

Kekistan has not only a fictional flag but an anthem, the 1980s pop song *Shadilay*. Its 'religion' is supposed to be worship of the cartoon character Pepe the Frog, transmuted into the Egyptian frog-god Kek. These symbols, along with hundreds of memetic variations, are used as a subcultural code on right-wing internet sites – for example, when white supremacists declare 'Kekistani' as a fictional nationality or Kek as a religion on census forms. At the same time, Kekistan is more than a code. It is what literary theory calls a 'conceit' – an extended metaphor with its own complex internal logic, designed to amuse.

Once you understand the logic, it is not very amusing.

The white supremacist and neo-Nazi movements in America never went away. But they were small. What's new is the emergence online of a widespread, though fragmentary, far-right culture among conservative-minded young people. Using bulletin boards such as 4chan, video channels on YouTube and a network of influencers on Facebook and Twitter, they have created a shared mindspace which stretches from overt fascist groups to the fringes of the Trump administration. Kekistan is the name of that space.

The Kekistan meme is evidence that a new rationale for fascism has emerged: a new form of techno-conservatism, opposed to the rights of women and ethnic minorities and based overtly on anti-humanist principles. Its danger lies not in its ability to mobilize a few thousand fascist activists of the old style. It lies in its ability to create synergies between three sections of the right that political science assumed over the past thirty years usually worked against each other: the extreme right (the overt fascists), the populist radical right, who tended to avoid violence and build their electoral base with appeals to nostalgia and cultural insecurity among working-class people, and mainstream conservatism itself.[3]

After the Brexit referendum and Trump's victory – both of which saw ethnic nationalist movements capture the mainstream – much of the academic sociology studying these movements will have to be revised. But in the meantime such movements are making new facts on the ground: the Salvini-led government in Italy, the far-right coalition in Austria and the high vote for the far-right Sweden Democrats in September 2018 show the Trump presidency added momentum and energy to right-wing authoritarian movements across the developed world.

Why did the breakdown of a coherent ideology justifying free markets and American power lead to the widespread adoption of ideas promoting racial and biological supremacy in the neoliberal heartlands? The road to Kekistan begins in the aftermath of the financial crisis.

The events of 2008 threw traditional conservatism into disarray, not just in the USA but across the world. In April 2009 a Koch-funded, ultra-conservative think-tank, the Cato Institute, convened an online seminar for right-wing intellectuals, entitled 'From Scratch'. The seminar's outcome was the Neo-Reactionary Movement (dubbed NRx in online forums), whose central conclusion was that the right should abandon democracy. Peter Thiel, the dotcom billionaire and founder of PayPal, outlined the logic. Even in the 1990s it was clear that 'capitalism is not that popular with the crowd', he wrote. Now, with massive state intervention to save the banks, it was impossible to imagine any American electorate voting to shrink the state, given the mass bankruptcies of both firms and savers this would bring. Thiel concluded: 'I no longer believe that freedom and democracy are compatible.'[4]

Thiel spelled out three ways to create new realities outside the democratic institutions. Two were fanciful: colonizing the sea and outer space. A third was not fanciful: to move politics into cyberspace where the normal rules do not apply. Thiel called for the emergence of a 'new world currency free from all government control and dilution', and hoped that platforms like Facebook would 'create the space for new modes of dissent and new ways to form communities not bounded by nation states'. Thiel's was implicitly a project to live 'despite' the capitalism of state bailouts and busted banks; to create online movements from below; to refuse the logic of the post-bailout reality and avoid participation in the official democratic process.

Meanwhile, the US computer scientist Curtis Yarvin, writing under the pseudonym Mencius Moldbug, had begun advocating replacing democracy with authoritarian rule. Yarvin/Moldbug would become the unofficial prophet of the neo-reactionaries, establishing a cult following with his rambling, assertion-ridden 5,000-word essays. Moldbug advocated that democracy should be replaced by a benign dictatorship modelled on a company, in which the figurehead has the right to milk the system for his own family's needs, as long as he guarantees economic freedom. The closest historical model was the same one Hegel had worshipped: the Prussian monarchy, only under an earlier monarch – Frederick the Great. Citing Hong Kong, Singapore and Dubai as successful twenty-first-century autocracies, Yarvin pointed out: 'They are weak only in political freedom, and political freedom is unimportant by definition when government is stable and effective.'[5]

Yarvin's work would become the justification for a new ultra-right political strategy after the financial crisis: stop wasting time on policy formulation; start creating power dynamics on the ground; promote cryptocurrencies to escape from the diktats of central banks and prepare the conditions in which a one-family kleptocracy can take power. Though they could not say so in public, the tech utopians of the right had reinvented the thought-architecture of fascism.

Though the tens of thousands of meme-sharers, rape fantasists and white supremacists probably never read these texts, enough of their content was distilled into simpler messages – via YouTubers and internet TV shows – to create a rich verbal and visual culture through which the NRx philosophy was communicated. Those, like left-wing writer Angela Nagle, who see the alt-right movement as the simple product of exasperation with campus political correctness, miss the fact that it was theorized in advance using the same resources – the Koch-funded think-tanks – that put Trump in power.

As early as the G20's Pittsburgh summit in 2009, I reported that the overarching fantasy of the 're-set' had entered right-wing populist politics. One worried member of the Libertarian Party, himself protesting against Obama, told me that 'too many people on all sides are beginning to fantasise about some kind of showdown in America', a rerun of the American Civil War with AR-15 rifles.[6] But it was not yet clear how the antidemocratic turn among right-wing thinkers would stimulate a movement on the ground. No section of the ruling elite could give it overt support. Even a figure such as Glenn Beck, who had stirred up hatred via Fox News in the early Obama years, started to warn the Tea Party against being provoked into armed revolution.[7]

The move from NRx to a significant far-right street movement was catalysed by an issue most people had assumed was settled: women's rights. Through a series of massive coordinated online attacks on individual women, the theory and practice of the alt-right spilled out into the world of blood and fear.

It had been bubbling under before the crisis. In 2007 came the attack on the US game designer Kathy Sierra, who advocated moderating comments on web pages. Her home address was published, Photoshopped imagery of her being raped and murdered was published, forcing her to disappear from public life. In 2010 came the cyber-harassment of eleven-year-old American Jessi Slaughter, when thousands of teenagers and adults republished her address and urged her to kill herself.

These and other high-profile incidents of misogyny laid the basis for #Gamergate, a large scale coordinated attack on feminist critics of sexism in the games industry, beginning in 2014, involving relentless attempts to drive them out of public life and towards suicide. One of its victims, Anita Sarkeesian, a journalist who had critiqued the macho culture in video games, was subjected to massively shared online rape threats.

While only a few thousand people read Cato Institute blogs, and maybe tens of thousands participated in the organized 'doxxing' (publishing the address and personal details of your target) and flame wars against individual women, tens of millions of people play computer games, and they are mainly young men. The organizers of #Gamergate consciously exploited the infrastructure of online gaming, through which hundreds of players at a time are randomly connected through voice servers.[8] There is no public record of actual in-game audio chats, but widespread anecdotal evidence suggests that this world became pervaded with #Gamergate propaganda.

The #Gamergate scandal was the catalyst that brought together the separate forces that would form the alt-right: 4chan users, professional trolls, men's rights activists, the traditional antifeminists of the Evangelical right and the ultra-right media group Breitbart News – propelling its star writer Milo Yiannopoulos to instant notoriety.

#Gamergate fitted the template the NRx writers had outlined: action outside politics to create new power dynamics. And in turn it created a new tactical template: target a victim with violent threats, frame the attack as a defence of your own right to free speech, use the constitutional issue as a catalyst to build mass support in the real world from other 'free speech-ers' and drive traffic to Breitbart News. Breitbart then forces the controversy into the mainstream media – asking 'why aren't they reporting this?' – following which a new breed of Fox News talking heads would proceed to normalize the attack and justify the victimization. Through #Gamergate the popular cultural language of the alt-right – cuck, SJW and feminazi – became acceptable among millions of conservative-minded young men, together with the tactic of 'gaslighting' (consciously attempting to manipulate your victim psychologically) and doxxing.

Why did the attack on feminism become the conduit for transmitting the theories and strategies of the alt-right into the consciousness of hundreds of thousands of young people? The most obvious answer lies in the reversal of male biological power, through birth control and equal rights legislation, that took place during the last decades of the twentieth century. Beginning in the 1960s, it is the most significant change in power relationships in human history. Capitalism adapted to it, indeed gained dynamism from it, by mobilizing women into the workforce, automating a large amount of the domestic work they traditionally did, turning the liberated sexuality of young women into a consumer brand and building whole new industries based on women's independent spending power.

But in 2008 the neoliberal route to women's empowerment was thrown into crisis. Rising female participation in the workforce, and a narrowing gender pay gap, looked benign to the sexist American man as long as the economy was expanding. After 2008, laws giving women formal equality at work and in family disputes, and protecting them from sexual assault – combined with the cultural normalization of female sexual independence – began to be reframed by the right as 'anti-male'.

If we list the obsessions of the men's rights activists (MRAs), it becomes clear that antifeminism is not peripheral to the alt-right: it informs its entire critique of the modern world. First, they claim that straight white men are uniquely the victims of the emergence of identity politics: while women, LGBT people and ethnic minorities have clear identities, giving them formal and informal rights, goes the argument, straight white men do not. In response, they have constructed their own oppressed identity: the 'Beta male', the young man who cannot find a sexual partner because women are too busy having sex with so-called 'Alpha males'.

It should go without saying that the whole thing is a laughably adolescent take on heterosexuality. But it is simply one of neoliberalism's fundamental tenets pushed to its logical extreme: that human beings are biologically unequal and that the market will reflect such inequalities by rewarding the strongest with success.

The 'Beta rebellion' is a conscious adaptation of an idea outlined by the nineteenth-century German philosopher Friedrich Nietzsche, in his attack on Hegel's concept of the 'end of history'. If it ever happened, said Nietzsche, the end of history would for most people mean a descent into powerless inactivity. The survivors would become 'men without chests' – i.e. weaklings, or 'last men' as Nietzsche called them. Because it accepts biological power as normal, the Beta ideology never questions why Alpha males should have power: that Alphas should run society is to them as natural as the best surfer always getting the first chance at the breaking wave. What's 'unnatural' to them is the new power of women to choose their partners and live their lives without sexist bullshit.

Nietzsche – not Hegel – became the neoliberals' philosophical hero because he justified the triumph of willpower: of the strong over the weak, and of the purposeful liar over the moral and ethical person. While neoliberalism worked, and while the global system sustained it, the elites were always obliged to hide their commitment to this 'Triumph of the Will' doctrine behind the façade of philanthropy and civilized discourse. Once neoliberalism hits crisis, then there is no more need to hide the belief in biological hierarchy.

If you accept that your own position as a man is biologically pre-determined, as do many of the foot-soldiers of the alt-right, then it's easy to see feminism as an attack on the biological order of things. As to strategies, for the Betas there are basically three: to imitate an Alpha male – by following the numerous guides to 'pick up artistry', in which women are tricked into having sex; to become an 'incel' – involuntarily celibate and at war with liberated women; or to detach from the world of heterosexual relationships into systematic porn use or voluntary celibacy, while waging a cultural war on feminism.

This is background to the emergence of the Proud Boys movement in America – a combination of fraternity-style cult behaviour combined with misogynist and anti-Muslim propaganda, fronted by an increasingly violent 'defence' arm called the Fraternal Order of Alt Knights. All told there are, at time of writing, maybe 6,000 members of the Proud Boys and the numbers of active neo-Nazis probably reach low five figures. But the ideological movement they are part of is big. The RedPill sub-Reddit – a popular site for antifeminism on the Reddit bulletin board system – has 226,000 followers; 4chan, the alt-right bulletin board of choice, boasts 11 million monthly users in the USA alone.[9] The wider problem is that – in the few short years since social media became a global reality – violent misogyny has become a pervasive subcultural identifier for the far right worldwide.

Coinciding with the sudden availability of free, high-definition online porn videos via broadband and 4G, the new misogyny mapped easily onto its prevailing storylines: the gang-bang, the submissive woman, the drunk woman tricked into sex, the black male as a sex-ual predator and the 'cuckold' – the Beta male forced to watch as more Alpha males (often black) have sex with 'his' woman. In 2017,

81 million people per day visited Pornhub, the most popular porn site on the internet, one third of them under the age of thirty-five.[10]

There is a familiar pattern here. The violent misogyny of the alt-right draws on a set of ancient prejudices given a new economic content and shaped into a new victim mentality by the far right in conditions of economic stress. Hannah Arendt, surveying the rise of the Nazis, warned it would be a mistake to confuse modern anti-Semitism for its medieval form. This bears repeating in relation to misogyny today.

After 2008, it is arguable that misogyny began to function as an ideological magnet for all other discontents, with its own symbolic language. Just as the fascists of the 1920s had accused Jews of spreading Cultural Bolshevism, the neo-reactionaries pinned the Cultural Marxist label on feminists – or 'feminazis', as they were labelled. The word 'cuck-servative', another staple insult of the alt-right, denotes the powerless conservatism of the business elite, which they deem responsible for the eclipse of male white power and the victory of Obama.

But the alt-right and the wider ethnic nationalist movement it draws from differ in an important way from the ideologies that fed fascism in the 1930s: they have a conscious and sophisticated theory of ideology itself.

In the 1999 blockbuster movie *The Matrix*, the protagonist, Neo, is trapped in a virtual reality constructed by his oppressors. In a critical scene he is given a choice – take a blue pill and stay in the fake reality or take a red pill and see the surrounding fakeness for what it is, at the cost of being deprived of happiness for ever and pitched into a permanent state of revolt. Stories with alternative reality themes have proliferated during the past thirty years – for example *The Truman Show*, *Westworld* and *Inception* – as a general metaphor for our inability to escape neoliberalism, or think beyond it. So when 'to redpill' entered the Urban Dictionary as a verb in 2004, it was initially politically neutral.[11] It meant achieving, or being taught to achieve, political consciousness against the prevailing ideology.

But by the time the RedPill sub-Reddit was launched in 2012, the concept of 'redpilling' had been completely colonized by the right.

The neo-reactionary critique of neoliberalism is that it has become too egalitarian, too rational, too democratic and too attached to 'universal' values. Yarvin/Moldbug described the power structures of the neoliberal era as 'the Cathedral': an unchallengeable thought-architecture in which universities and the press – that is, our basic sources of rationality and truth – practise 'comprehensive thought control arrayed in defense of universalistic dogma'.[12]

While left-wing theories of ideology stress the way illusions about reality arise from our lived experience, for the neo-right the thought control is always consciously imposed by their enemy and therefore more easily escapable. All you have to do is take the red pill and 'wake up'. Who supplies the pill? Rupert Murdoch via Fox News, Robert Mercer via Breitbart and tens of thousands of proselytizing Betas pumping out transgressive thoughts, words and calls to action via the bulletin boards and Twitter, activity known as 'shitposting'.

By 2014 the American right had reassembled the fragments of the old conservatism around: (a) a fully theorized rejection of democracy; (b) violent misogyny as the main driver for its victim narrative; (c) hostility to rationalism, the universities and the media; (d) rejection of the concept of universal human rights; (e) the stratification of humanity according to biological differences: ethnicity, gender and IQ. Around this backbone, you could arrange your pet obsessions: anti-Semitism for some, anti-Islam for others, anti-gun control for almost everyone.

All the alt-right needed was an outside force to aid it and an internal enemy to fight. The internal enemy was easy enough to anticipate: America's black, Muslim and Hispanic populations. The outside force – in a bizarre inversion of the Cold War narrative – would be Putin's Russia.

The year 2008 saw not only the collapse of the world financial order but, barely noticed, the first significant break in the global diplomatic order. In August that year, Russia invaded and defeated the small Black Sea state of Georgia. Despite condemnation by every one of the multilateral institutions set up after 1989, Putin had – at very little cost – prevented the expansion of NATO into the Black Sea and established a stronger footprint for his own forces there. We were slow to understand it, but after Georgia the world was effectively multipolar once again. But this time with an important difference.

At university in the late 1970s, I attended seminars by a seasoned Kremlin-watcher. His office walls were papered floor–to-ceiling with 10x8 mugshots of Soviet bureaucrats, the power structure inside the Kremlin carefully mapped and annotated in a mountain of box files. To predict Soviet behaviour during the Cold War, you had to grapple with a complex, highly educated bureaucracy, working with explicit doctrines and limited by known, institutional checks and balances.

By 2008 my old professor would have needed only one mugshot, while in place of the box files he would have needed only a psychiatrist's report on Vladimir Putin. Instead of dealing with a complex bureaucracy executing a collective doctrine, the West was now confronted by a single man prone to appearing bare-chested on horseback – his reactions to all world events filtered via a daily briefing from people who dare not displease him. Back then, you could study the entire command structure of the Kremlin and its secret state, but what mattered now was in Putin's brain.[13]

After Georgia, Russia's temporary president, Dmitri Medvedev, spelled out Russia's new foreign policy on Putin's behalf. The world is now multipolar. Russia will resist any attempt to encroach on its territory. It will protect Russian speakers everywhere and defend its 'privileged interests' in certain regions.[14]

What Medvedev outlined was the so-called 'Great Power' doctrine, and the implications of his speech should have been clear. Russia's aim was to weaken the multilateral institutions underpinning the global order (the UN, the Organization for Security and Co-operation in Europe – OSCE – and various treaties on disarmament) and to promote instead a world order based on direct horse-trading between itself, the USA and China. Central Asia and the Black Sea would be Russian spheres of influence, with Ukraine pulled back into the Russian orbit after the victory of Viktor Yanukovych, the pro-Kremlin candidate, in its 2010 elections – aided and abetted by Trump associate Paul Manafort.

Russia would militarize its portion of the Arctic and maintain both Iran, Lebanon and Syria as proxies in the Middle East. Through 'hybrid warfare' in Europe, encouraging far left and far right parties and various nationalisms, it would weaken in practice the commitment of EU members to NATO's collective defence doctrine. It wanted, in summary, a sizeable piece of the world as its backyard.

Western security experts had no problem reading Putin's intent. But it did not fit into their worldview, which told them that globalization would steadily deepen, and that free markets would drive democracy. To Western strategists, Putin and the clique of former intelligence and military officers who surrounded him, known as the *siloviki*, were seen merely as authoritarian caretakers, destined to fade away once the new, Westernized middle class could summon the political maturity to take control. As far as Russia's Great Power doctrine was real, they assumed, it was primarily defensive, not offensive.

These were the assumptions that led the Obama administration into its strategic and ultimately disastrous 'pivot' towards Asia. Hillary Clinton, then secretary of state, spelled out the logic in her November 2011 article 'America's Pacific Century'.[15] If globalization was going to continue, she said, the USA had to shape the flow of goods and services in East Asia towards its own interests. It had to shore up its military alliances with Japan and South Korea and prevent China becoming the regionally dominant naval power, with the ability to dominate the world's most vital sea lane, the South China Sea. Pivoting to Asia meant leaving Europe to deal with any threat from Russia.

The timing was disastrous. One month after Clinton's article hit the newsstands, networked protests broke out in Moscow. The old deal – Putin delivers prosperity and the modern, liberal, networked section of society stays out of politics – seemed in question. To smash the protest movement, Putin not only arrested its leaders and mobilized nationalist right-wing thugs, he passed a law introducing a state blacklist of internet sites.[16] Then he passed a law designating 148 NGOs broadly aligned with democratic goals as 'foreign agents' and ramped up state harassment of foreign media.[17] In Putin's view, the upsurge of democratic protest in Russia – and its puppet Syria – had been created by the West. In response, Putin tasked his security apparatus to begin designing a different kind of response – one for which Clinton's State Department, now obsessed with Asia, looked totally unprepared.

In 2012 Russia unilaterally deployed its troops into Syria. After the pro-Moscow regime in Kiev was overthrown in 2014, Russia responded by invading and annexing Crimea, and then starting the civil war that would leave two Russian-speaking provinces of Ukraine partitioned and occupied by Putin's armed forces.

Only as this game unfolded, move by move, did it become clear that Putin was actually waging an offensive hybrid war. Hybrid warfare has been defined as a mixture of 'conventional weapons, irregular tactics, terrorism, and criminal behaviour in the same time and battle space' to obtain political objectives.[18] Democracies cannot, by definition, prosecute it comfortably.

Central to the Russian doctrine, which the Kremlin calls 'new generation warfare', is information. Conventional armies fight an information battle against opposing commanders; disguising movements, jamming communications, waging psychological warfare on civilian populations – all to shape outcomes in the physical battlespace of combat vehicles, aircraft and ships. Hybrid warfare is different. As one NATO analyst noted, 'the Russian view of modern warfare is based on the idea that the main battle space is in the mind'.[19]

It is likely that even now – after Putin has installed his preferred candidate in the White House, influenced the Brexit vote, spent €20 million backing the Front National in France – many people in the West do not realize that Russia is waging a war inside their minds. Populations with democratic cultures are conditioned to assume that the rule of law exists, and that those who break it are liable to be indicted. Yet in fact, large parts of the world are already a rule-free environment, where might and intelligence can achieve any outcome desired by the super-rich.

However, Putin's new strategy is not just a problem for Western governments. It is a problem for everyone who wants to replace those governments with a radical alternative to neoliberalism.

In the 1930s, democrats and socialists were fighting fascist movements largely hostile to the nationalisms of other countries. Today, fascism, ethnic nationalism and authoritarian regimes are prepared to inter-operate and support each other. And all right-wing movements capable of destabilizing Western democracy, or weakening NATO and the European Union, can rely on tacit or even outright support from the Kremlin – above all in the sphere of networked disinformation.

In 2011, amid the euphoria of the Arab Spring and the Occupy movements, it looked as though networked communications had made all propaganda instantly capable of being refuted. Governments had lost their power to control the imagery of war and conflict; the official version of events could be checked with on-site witnesses in real time. Right-wing and authoritarian governments could survive by creating an information bubble among their supporters – but that would never be enough to maintain consent.

Part of their solution was fake news: the manufacture of stories that are patently untrue. Another part of the strategy, pioneered in Russia, was to pollute the networked space with so much disinformation and abuse that people recoiled from it. By 2013 Russian 'web brigades' – young people operating out of 'troll houses' – were adept at poisoning the atmosphere of any comment forum. Reporters found one paid troll posting comments such as 'Navalny is the Hitler of Our Time' – an attack on the main leader of the 2011 protest movement – for $36.50 per eight-hour shift.[20]

Soon the effort would become automated. By 2015, after the murder of oppositionist Boris Nemtsov on a bridge overlooked by the Kremlin, hundreds of Twitter users began simultaneously posting claims that Nemtsov had been killed by Ukrainians because 'he stole their girlfriend'. Researchers later identified 17,590 Twitter accounts spewing out these attacks: apart from making an average 2,800 Tweets each, they had barely interacted with anyone else on the network. They were machine-controlled 'bots'.[21]

It is important to understand how the trolls, the bots and the fake news creators work together as a system. Their aim is not to convince others that their version of reality is true; it is to make the atmosphere of online political debate so aggressive and unpleasant that ordinary people shrink from it; to raise the possibility that all sides are engaged in a propaganda war, and that therefore no news is trustworthy.

But Vladimir Putin's trolls and the willing purveyors of disinformation in the Western far right are only the supply side of this fake news economy. The bigger problem is the demand side.

*

The #BlackLivesMatter movement started in July 2013 in protest at the acquittal of neighbourhood watch coordinator George Zimmerman, who had shot dead black teenager Trayvon Martin during an altercation in Sanford, Florida. The black women who coined the hashtag and spread it were not only well versed in the techniques of networked activism but came from a generation of educated, upwardly mobile black people. What in 2011 had been new and experimental to the mainly white, middle-class young people in Zuccotti Park was now a coherent methodology which could be taught and learned.[22]

What moved #BLM from a defensive protest to an offensive and dynamic campaign for human rights was the 2014 uprising in Ferguson, Missouri, following the police murder of Michael Brown. This prompted a military-style occupation of the city by armed police. Hundreds of African American protesters held their hands up, chanting at the armed cops 'don't shoot'. Amid repeated killings of unarmed black Americans by the police and law enforcement auxiliaries, this was not mere irony.

After Ferguson, tens of thousands of young people from the most educated, literate and articulate generation black America has ever produced took up activism around the objectives of #BlackLivesMatter. Using techniques pioneered in the Occupy movement three years before, and channelling the knowledge accumulated in years of studying the civil rights movements of the 1950s and 60s, #BLM launched a structural challenge to the way black oppression had been shaped during the free-market era.

Neoliberalism had inflicted decades of poverty and criminalization on America's black communities. The consequence – because of the USA's draconian laws depriving felons of the vote – was widespread voter disenfranchisement. President Obama, though always ready with sympathy for the victims of police killings, had done little to alleviate these structural underpinnings of black oppression. By challenging this structure, #BlackLivesMatter triggered the deepest residual nightmares of white racism.

Historically, every victory for black human rights in America had been followed by an economic and social defeat. When slavery was abolished after 1865, the poor blacks of the South were turned into sharecroppers, oppressed by segregation laws and terrorized by lynchings under the Jim Crow system of apartheid. After the 1964 Civil Rights Act, a combination of geographic segregation and economic crisis turned many among the black urban poor into an underclass: denied education, decent jobs, routinely incarcerated and deprived of voting rights.

But what happens if the law enforcement and prison systems can no longer be weaponized to enact the oppression of black people? What happens if the police are forced to stop terrorizing random black car drivers and those merely standing in their own doorways at night? And if the courts are forced to stop sending black men to work as semi-slaves in privately run jails?

The answer is an unprecedented challenge to what the black sociologist W. E. B. Dubois dubbed in the 1930s the 'public and psychological wage' of whiteness. If you were white, no matter how poor, you had a better chance of rising out of poverty and very little chance of finding yourself killed in gang violence or by the police, or incarcerated. As the ideology of neoliberalism spread, boosting the idea that poor people 'deserve' their fate as unsuccessful competitors in an efficient market, this added a technocratic rationale for black poverty and oppression. You could reject, as most conservatives did, the so-called 'race science' contained in Charles Murray's 1994 book *The Bell Curve,* but still accept there was something lawful about the poor economic outcomes for racial minorities in the USA.

As the political scientist Joel Olson pointed out, by 2008 legal equality for black people masked 'a system of tacit and concealed racial privileges that is reproduced less through overt forms of discrimination than through market forces, cultural habits, and other everyday practices that presume ... white advantage is the natural outcome of market forces and individual choices'.[23]

This in turn had found its political expression in the Republican strategy of racializing mainstream politics. 'The Democrats are a part of the white elite who pander to the black criminal underclass' would be a fair summary of the subtext all Republicans, mainstream or not, were prepared to propagate at election time.

The ultimate 'tacit and concealed racial privilege' was that a white person could call the cops on a black person and put them in fear of being violently assaulted, unfairly jailed or, in extremis, shot dead. By challenging that – and doing so on the basis of social theory and constitutional legality, not religion or emotion – the founders of #BLM threatened to smash the 'glass floor' that separates all black Americans from poor and insecure white Americans.

And that's a big thing. Fear of black liberation extends far beyond the confines of the alt-right, into the heart of the white middle class. Witness the numerous 'Permit Patty' incidents recorded on social media, where white racists call the police on black people for lighting barbecues, bathing in certain pools or, in the actual case of 'Patty' herself, against a black child selling bottled water to football fans as they passed outside her apartment.[24]

Once #BLM became both a movement and a consciousness among young, networked, black Americans, spilling over into the kneeling protests by black NFL players during the US national anthem, reaction to it enabled the alt-right to further colonize the American conservative agenda.

'Race science', having lain dormant and largely discredited on the fringes of right-wing thought, surged back into the limelight. Murray himself has been fêted by the alt-right campus circuit, while websites like American Renaissance agglomerate all available pseudoscience devoted to proving black people are genetically less intelligent than whites. And this is no accident – for at the core of the alt-right's thinking about race, just as with gender, is its opposition to the universality of the human existence, and its belief in the biological lawfulness of inequality.

Since the 1990s, when right-wing populist groups began to make electoral gains in Europe, political scientists have tried to understand the grievances that drive them, and the ideologies they create. Their unspoken aim was always to work out ways in which the new far right could be contained and prevented from joining up with the smaller remnants of outright fascism.

Controversy raged, but a fair summary of the consensus was that for the outright fascists the main grievances were economic, while for the right-wing populists the grievances were cultural, driven by a perceived loss of status among existing working-class communities faced with migration. Both groups were effectively 'victims of modernity' who, it might be hoped, would at some point accept the loss of well-paid jobs, social cohesion and ethnic privilege. The fascist groups tended to be statist and highly socially conservative, while the populist parties favoured free-market economics and were prepared to co-opt gay rights and women's rights (for example against female genital mutilation) as issues with which to stigmatize migrant groups, particularly Muslims.

Because of the time lag, the majority of academic research done on the new far right precedes both the 2008 crisis and the election of Trump. But the new dynamics we are dealing with are pretty obvious at micro level.

Take my home town, Leigh, in the northwest of England: the almost immediate outcome of the 2008 crisis there was to bring the fascist British National Party into electoral politics. In the 2010 general election its candidate came from nowhere to gain 2,700 votes, or 6 per cent, beating the (also new) right-wing populist party UKIP, which obtained 1,500. This was despite the fact that the BNP had no openly functioning group in the town. By the 2015 general election, UKIP was on 9,000 votes, or 20 per cent, having absorbed all the fascist votes and up to 4,000 taken directly from Labour and the Conservatives. In the Brexit referendum of 2016 the town voted 2:1 to leave the European Union, fulfilling the BNP and UKIP programme. In the June 2017 election UKIP's vote promptly collapsed back to 2,700, with the Conservatives adding 6,000 to their previous total.[25]

In short, far-right politics is a work in progress. It is successful when it has a single major grievance around which to polarize the electorate; the smaller, hardcore fascist group is always prepared to gravitate towards the successful populist party, and a lot depends on how mainstream parties react.

Across Europe, the right-wing populist parties were learning how to outmanoeuvre the centrists, themselves paralysed by their allegiance to an economic system that didn't work. The result is the present European landscape: right-wing populist parties rule Poland, Hungary, the Czech Republic and – via coalitions – Austria and Italy. And as they gain state power, or the parliamentary privileges that come with double digit electoral scores, they do what Trump has done: use the machinery of the democratic state to legitimize hate speech, paralyse law enforcement against outright fascists and target migrant communities for repression and deportation.

Valuable though it is, the slow-moving world of academic studies into the far right is struggling to catch up with the dynamic and unpredictable reality. What drives people to the far right is no longer simply economic or cultural insecurity, it is the fact of seeing such parties legitimized, in power, actively dismantling liberal democracies from above.

In the 1930s, confronted with the swing towards fascism among the working poor, left sociologists and psychiatrists developed the theory of the 'authoritarian personality'. Some of the earliest research was conducted by the Marxist sociologist Erich Fromm through a questionnaire that in 1929 he gave to 584 working-class adherents of fascism, social democracy and communism. Fromm's conclusion was that within the German left there had, during the good years, existed people whose 'basic personality traits' were at odds with their political alignment. This group resented the elite, but values 'such as freedom and equality had not the slightest attraction for them, since they willingly obeyed every powerful authority they admired'. As the social crisis of the early 1930s intensified, this group, Fromm concluded, were transformed from unreliable leftists into convinced Nazis.[26]

In the 1950s, building on Fromm's work, Theodore Adorno led a team of US-based psychiatrists who claimed to be able to measure the propensity to fascism of individuals according to their attitudes to authority, family, homosexuality, race and sex, describing the personality of the typical fascist recruit as the 'authoritarian rebel'. In addition to being slammed for inconsistent methodology, by the 1960s Adorno's work seemed irrelevant. Fascism had evaporated, white supremacism as a doctrine was underground; the personality type the establishment feared was distinctly anti-authoritarian.

Today, however, we are faced once again with the problem of where mass fascist psychology comes from. The rising authoritarian nationalism of Trump, his cohorts and imitators inhabits a common ideological space with overt right-wing racism of the Breitbart variety and outright neofascism. Though it is masked within metaphorical spaces such as Kekistan or the 'chan' bulletin boards, where nothing is apparently taken seriously, the anger and outrage stoked online repeatedly spills into violent reality.

For progressives the task is not only to defeat Trump, and to drive his authoritarian nationalist counterparts in Europe from office at the ballot box. It is also to prevent the evolution of authoritarian nationalism towards fascism; to break up the 'temporary alliance of the elite and mob' before it achieves the permanent destruction of democracies, constitutions and the global order.

For this it is necessary to understand what specifically characterizes the alt-right mentality. At time of writing only one evidence-based study of the alt-right has been completed, and its results are stark. Patrick Forscher and Nour Kteily surveyed 447 Americans identifying themselves as alt-right, comparing them with a slightly smaller group drawn from the general population.

Like Fromm, they tried to establish 'personality traits', looking in particular for the so-called 'dark triad' of narcissism, Machiavellianism and psychopathy. They also wanted to see if, as with Fromm's Nazis, the alt-rightists were attracted to authority. In the end, however, though the alt-right respondents were slightly more authoritarian and slightly more prone to psychopathy and narcissism, the most startling difference was in their willingness to label their opponents inhuman.

Asked to place various groups along an evolutionary diagram ranging from ape to upright man, the alt-right people consistently dehumanized 'Arabs, Muslims, Hispanics, black people'. Collectively, the alt-right people placed these groups as around halfway along the evolutionary scale between chimps and *Homo sapiens*. Asked to place 'feminists, Democrats, Republicans who refuse to vote for Trump and journalists' on the same scale, they awarded the same score. The only other significant divergence between the neofascist mentality and the norm was when it came to reporting their own propensity to violence and online harassment, which they readily admitted, and their trust in the mainstream media, which was non-existent.[27]

When researchers broke down the responses within the alt-right group, they found another startling result: about half were 'supremacists', prepared to rate black people, feminists and Hispanics at a level just above that of the chimpanzee; while a group they labelled 'populists' tended to rate these groups simply as 'subhuman'. And the two subgroups were different in their attitude to violence. Basically, the supremacists – numbering around half of those surveyed – revelled in their violent acts towards political opponents.

At one level this survey only confirms what survivors of the Holocaust have been telling us for decades: that the ability to dehumanize an ethnic group legitimizes violence against them. But it also confirms the centrality of the biological power thesis for right-wing thought. It is anti-universalism, not the tendency to worship those with authority or sociopathic tendencies, which is the standout marker for the twenty-first-century fascist mentality.

Not all those prepared to segment the population into real humans and apes become fascists. But the recognition of our universal humanity – uniting differences in skin colour, face shape, religion and culture – is the defence line for preventing the slide towards both right-wing authoritarianism and full-blown fascism. Once again, the defence of the concept of the human being, with universal rights, is key to resisting the slide to chaos.

Nobody designed this catastrophe. It was caused, if anything, by the Western elite's disdain for the rational design of societies. Now this threefold crisis – strategic economic stagnation, global fragmentation and the rise of irrationalism – characterizes and dominates the age we live in.

We have seen above how neoliberalism hollowed out our concept of the human being; how the performative nature of everyday life allowed stinking prejudices to fester behind the smiles and 'have-a-nice-days' required by corporate etiquette; how the crisis of the neoliberal system led sections of the elite to abandon globalism and the commitment to democratic values, and how the thought-architecture of fascism was rediscovered via the gaming laptops of frustrated young men and the meanderings of discredited pseudoscience.

Now it is time to understand what a catastrophic moment in human culture this might create. If globalization falls apart, that's the end of a forty-year process. If technology and productivity have ceased to drive growth, that would be the end of a 200-year trend. But if the human-centred thought patterns, norms and behaviours which underpin democracy are actively rejected by millions of people, that is a much bigger reversal.

The emergence of widespread, popular anti-humanism does not just hold open the door for some fascists with stupid flags. It opens the door for our surrender to machine control – and, in the face of it, the ordinary humanism of the liberal mainstream has begun to falter.

7

Reading Arendt is Not Enough

Trump's victory was for many people a shock moment. It dramatized how close we have moved towards the return of totalitarianism, how prevalent theories of white and male supremacy have become, and how fragile the truth. Like victims in a vampire movie, we grabbed the garlic closest to hand: books by revered humanist writers from the 1940s and 50s, telling us how to resist.

The writings of George Orwell and Auschwitz survivor Primo Levi flew off the shelves. The novels of repentant communist Arthur Koestler and the persecuted Soviet journalist Vasily Grossman were revived. Above all, the work of the German-born political philosopher Hannah Arendt gained massive popularity. In the months after Trump came to power Arendt became something like the patron saint of liberal angst. Like all these writers she had been in the 1940s and 50s part of the humanist reaction to the experience of Nazism, the Holocaust and the Cold War.

In 1951 Arendt wrote that the ideal subject of a totalitarian state is not the convinced Nazi or communist but 'people for whom the distinction between fact and fiction (i.e., the reality of experience) and the distinction between true and false (i.e., the standards of thought) no longer exist'.[1]

This was a near-perfect description, sixty-five years in advance, of the electorate shaped by Trump's rallies, Fox News and the Kremlin's secret Facebook ads. What had made people susceptible to fake news in the 1930s, Arendt argued, was loneliness: 'the experience of not belonging to the world at all, which is among the most radical and desperate experiences of man'.[2] That's the kind of loneliness you experience today in small-town USA, or in the left-behind industrial towns of Britain, or the backwaters of Poland and Hungary – all heartlands of the new authoritarian racism. It's also, paradoxically, the kind of loneliness you can experience in a networked society: how many of the woman-hating and racist mass shooters in America are, after the event, described as 'loners'?

Arendt's study of how totalitarianism was spread via sympathizers inside democratic institutions and the mass media also resonates today. Through them, she argued, fascist movements 'can spread their propaganda in milder, more respectable forms, until the whole atmosphere is poisoned with totalitarian elements which are hardly recognizable as such but appear to be normal political reactions or opinions'.[3] Today's right-wing media ecosystem, through which the hardline fascists of the alt-right spread their lies, via the so-called 'alt-lite' websites such as Breitbart into the mainstream channels like Fox News, corresponds exactly to Arendt's description.

Later, in her report on the trial of the Nazi war criminal Adolf Eichmann, Arendt coined a phrase that could be applied to many of today's authoritarian kleptocrats: 'the banality of evil'. Thousands of Nazi functionaries like Eichmann had participated in mass killing, only to return home each evening to humdrum domestic life. What made them capable of this, Arendt argued, was the loss of their ability to think: 'The longer one listened to [Eichmann], the more obvious it became that his inability to speak was closely connected with an inability to think, namely, to think from the standpoint of somebody else.'[4]

This, in turn, was rooted in the modern bureaucratic lifestyle. Totalitarian states make people into cogs in an administrative machine, Arendt argued, 'dehumanizing them'. Worse, she said, this might even be a feature of all modern bureaucracies.

Finally, as discussed in Chapter 6, Arendt understood what it was that could bind together a 'temporary alliance of the elite and the mob': the realization that their ideologies would make sense only if they could reverse historical progress. Both needed 'access to history', Arendt argued, even at the price of destroying the society around them. Today, for both the millionaires surrounding Trump and the Betas marching by torchlight through Charlottesville, that is the aim: rewind history and destroy the global order.

Arendt, then, provides important insights even at half a century's distance. But after 1989, with the collapse of the Soviet Union, it seemed as though the spectre of totalitarian systems had gone for good. There were still dictatorships, but they were shabby affairs in countries too poor to support a Nazi-style bureaucracy, let alone practise systematic mind-control over their populations. By 2000, when the philosopher Tzvetan Todorov wrote the magnificent history of twentieth-century resistance *Hope and Memory*, he concluded: 'Totalitarianism now belongs to the past; that particular disease has been beaten.'[5]

As we watched the forces that brought Trump to power we understood that the totalitarian-minded people Arendt had described have returned. But why?

As we copied and pasted insights from Hannah Arendt into our Facebook pages, and held up her words on placards at anti-Trump rallies, some disturbing questions arose.

First: if a successful free-market democracy like the USA is capable of producing a Trump, doesn't that make this moment worse than the 1930s? Hitler and Stalin were the products of state-dominated economies that hit crisis; they led subservient and poorly educated populations, who had been trained by generations of factory work and military conscription to obey the hierarchy above them. Germany had experienced precisely fourteen years of constitutional democracy in the 200 years before Hitler; before Stalin, Russia had experienced precisely none. Early twenty-first-century America, on the other hand, is a society full of educated people and with an uninterrupted democratic tradition going back to 1776. For the US to produce a fascist-like mass movement and a kleptocratic attack on the constitution was not in Arendt's script.

Second: while the dictators of the 1930s did rely on blurring the distinction between truth and lies, they were greatly helped by their absolute monopoly on information, and indeed disinformation: the elite controlled the printing press and the state controlled the radio stations. Even the possession of typewriters was strictly controlled, both in the Third Reich and the Soviet Union.[6] No such monopoly on information exists today – so what made so many people fall for the fake news strategy?

Third: Hitler was destroyed by Stalin. The entire post-war world in which Arendt, Orwell, Koestler and Levi wrote their critiques of the totalitarian mindset was created by the victory of one totalitarian state over another. If the West is today under threat from a resurgent totalitarian impulse, where is the external force capable of smashing it, as the Allied and Soviet armies did in 1944–5?

Of all the anti-authoritarians of the 1940s and 50s, it is Arendt who evades these questions most skilfully. Orwell and Koestler fought fascism in Spain: Koestler as a card-carrying communist, Orwell as a member of the far-left POUM militia. Levi fought as a partisan in 1943, in a group allied to the liberal-socialist Partito d'Azione. Vasily Grossman, the first Soviet journalist into the remains of the Treblinka concentration camp, had served throughout the war as a Red Army journalist. Every one of them understood they were morally compromised by the antifascist war they had taken part in.

Levi's partisan unit disintegrated after they were forced to shoot two volunteers for indiscipline. Koestler's portrait of the ruthless Soviet commissar was based in part on his own actions as a Comintern spy. Grossman had denounced other writers and managed to report the Red Army's advance across Europe without public mention of its mass rapes and massacres. Orwell's poem, 'The Italian Soldier', about an anarchist volunteer in the Spanish Civil War dramatized the problem of fighting fascism in alliance with Stalinism. 'The lie that slew you is buried,' Orwell wrote, in a bitter eulogy to his presumed-dead comrade, 'under a deeper lie.'[7]

Each of these writers committed violence in the name of antifascism. In their work, antifascist violence is seen as inevitable, if tragic – and leads ultimately to the strengthening of Stalinism, bureaucracy or inhuman attitudes. Arendt committed no antifascist violence – though she was jailed for political opposition to the Nazis in 1933 and had to escape from France after they invaded, arriving in the USA in 1941.

Practically, Arendt solved the problem of fascism versus Stalinism by escaping to America, an achievement nobody could begrudge. Theoretically, however, she solved it by claiming that US constitutional democracy was a form of industrial society uniquely immune to totalitarianism. In her 1948 lecture to a socialist club in New York,

Arendt outlined a clear theory of American exceptionalism from totalitarian tendencies:

> The American Republic is the only political body based on the great eighteenth-century revolutions that has survived 150 years of industrialisation and capitalist development, that has been able to cope with the rise of the bourgeoisie, and that has withstood all temptations, despite strong and ugly racial prejudices in its society, to play the game of nationalist and imperialist politics.[8]

The USA, Arendt claimed, was a twentieth-century democracy which 'lives and thrives' by an eighteenth-century philosophy – that is, the utilitarian Protestant individualism written into the Constitution. The practical role of the philosopher was to improve US society by criticizing it – as she would do over black civil rights and Vietnam.

Arendt was a courageous opponent of tyranny, but instead of deifying her we should understand her ideas in their context. Nazism, she said, had emerged out of the 'vacuum resulting from an almost simultaneous breakdown of Europe's social and political structures'. When the Nazis said that the old order had collapsed they were, in this sense, simply 'lying the truth', she argued.

But Arendt never explained why Europe's social and political structures broke down. She preferred to describe innate tendencies towards evil – in the subterranean culture of anti-Semitism, or imperialist white supremacy – which 'crystallized' into Nazism and Stalinism. But crystallization is a physical process with cause and effect. If you are looking for an explanation of what caused the similarity between Nazism and Stalinism, look elsewhere: Arendt was a theorist of 'what's gone wrong and how should humans live?' – not 'what's happening and why?'

The fashionable claim that Arendt was the first person to identify the common features of the totalitarian projects of Nazism and Stalinism is ludicrous. Of all the people she mixed with, and whose work she would have read in America in the 1940s, she was the among the last to do so.

Throughout the 1920s, anarchists and socialists from the anti-Bolshevik tradition had warned that the Russian Revolution had the potential to create a dictatorship, mirroring the worst of what had happened in the West. When they considered the source of this danger they located it in the 'backwardness' of Russian society, or the uneducated level of the working class. When industrial-scale lying and oppression took off – with the ascendancy of Stalin's faction in 1927 – it was thinkers from the socialist and communist traditions who first proposed it might signal the emergence of something new, rooted in technological progress and the bureaucracy of modern states.

The Austrian socialist Lucien Laurat proposed in 1931 that the USSR was neither capitalist nor socialist, but 'bureau-technocratic': a new ruling caste had seized control and imposed a new form of class society. Laurat explicitly connected this to the emergence of managerial bureaucracy in Western countries, creating 'another form of exploitation of man by man' to replace capitalism.[9]

By 1937 the Soviet Union was practising industrial-scale murder. The Moscow show trials were merely the shop window for a vast purge that would, in the space of just two years, kill an estimated 1.2 million people – mainly left-wing communists, militant workers, political oppositionists and army officers deemed likely to side with them.[10]

It was in the aftermath of the Moscow trials that an oddball left-winger called Bruno Rizzi published a book entitled *The Bureaucratisation of the World*. In it, he argued the Soviet bureaucracy was simply a Russian expression of a new form of class society that was replacing capitalism all over the world: 'bureaucratic collectivism'. Rizzi proposed that in Russia, Germany and America this new bureaucracy had replaced the proletariat in driving historical progress. Both Nazi Germany and Mussolini's Italy, Rizzi said, had acquired an anticapitalist character: 'the social character of their countries is the same'.[11]

When Hitler and Stalin signed their peace pact in August 1939, dismembering Poland and leaving Germany free to wage war on Britain and France, Rizzi's 'bureaucratic collectivism' thesis took off powerfully inside the Western left. James Burnham, one of Trotsky's leading followers in the USA, declared the USSR, Nazi Germany and Roosevelt's America to be three kinds of 'a new form of exploitative society'. This 'managerial revolution' was destined to triumph everywhere, leaving historical progress with no option but to operate through the actions of totalitarian dictators. Compared to Arendt, whose *Origins of Totalitarianism* was criticized for being softer on Stalinism than Nazism, Burnham's theory was clear: the two are exact equivalents.

In George Orwell's masterpiece, *Nineteen Eighty-Four*, it is Burnham's ideas that are parodied in *The Book* – the secret manual of the underground movement trying to overthrow Big Brother. Orwell rejected Burnham's claim that the world was about to become three unmovable totalitarian dictatorships, but explored – as a warning – how it might come about: by suppressing all knowledge of the past; by turning language into political jargon so that people can't think rebellious thoughts; and by repressing sexual desire. Orwell's hero, Winston Smith, does find out about the past, does maintain a critical private language in his diary, and most certainly follows his sexual desires. But he is captured by the ingenuity of the Party, which has created a fake opposition leader, Emmanuel Goldstein, modelled half on Trotsky, half on Burnham, to entrap anybody who rebels.

These ideas – from Rizzi to Burnham to Orwell – had been current for more than ten years when Arendt wrote *The Origins of Totalitarianism*. What distinguished Arendt, then and later, was her refusal to explain why totalitarian ideologies triumph. 'There is an abyss,' she wrote, 'between men of brilliant and facile conceptions and men of brutal deeds and active bestiality, which no intellectual explanation is able to bridge.'[12]

If we are going to use Arendt as a guide for today, this conceptual void is a big problem. It is one thing to say that in the late 1920s the old European society collapsed and left a vacuum. The question that event posed is: why was that vacuum filled with such extremely similar ideologies and actions focused around inhumanity, death camps, organized lying, torture and the suppression of rational thought and language?

The missing idea in Arendt's thought was class. She correctly identified the brutalities of fascism as originating in those of late-nineteenth-century colonialism. She borrowed an idea from the Polish Marxist Rosa Luxemburg, that European states needed to export their excess savings and populations to their colonial possessions. She understood that imperialism created the material basis for an alliance between the 'elite and the mob', based on master-race theories; and that fascist movements were composed of people among both rich and poor whose interests suddenly converged on the collapse of the old order. She also was right to point out that the reformist socialists in pre-1914 Germany overlooked the dangers of working-class fascism because it didn't fit into their theories of class struggle.

But Arendt failed to understand the class dynamics of the societies that produced both fascism and Stalinism. The working-class revolts of the early twentieth century, and their failure, explain almost everything Arendt chooses not to explain about the rise of totalitarianism.

With fascism – in Italy, Germany and Spain – it is the inability of the capitalists to go on buying off a layer of workers, and the sheer size and social power of the radicalized labour movements, that obliges the elite to rely on militarized right-wing groups to smash the unions and the socialist parties. With Stalinism it is the backwardness of Russia, the isolation and atomization of the working class after three years of civil war, which by the mid-1920s allows a new class of bureaucrats to take the place of the old bourgeoisie. Unless you understand that working-class self-organization was the spectre haunting the European elite, from the global mass strike movement of 1911–13 right through to the defeat of fascism amid communist-led uprisings in 1943–4, you cannot understand why that elite became so prone to supporting fascism in the mid-twentieth century.

Today's events, however, pose questions that Arendt's methodology is even less suited to answer. Neoliberalism's collapse has stripped the current model of capitalism of all meaning and justification. Even in Arendt's beloved 'American Republic' the vacuum is being filled by an ideology hostile to human rights, to universalism, to gender and racial equality; an ideology that worships power, sees democracy as a sham and wishes for a catastrophic reset of the entire global order.

Worse, the number one weapon for the US right is that self-same 'eighteenth-century philosophy' which Arendt assumed had given Americans immunity from totalitarian rule: their individualism, which has been turned against them during thirty years of free-market rule, and their belief that economic choice constitutes freedom.

Arendt, in a phrase that still resonates, said that 'what the mob wanted . . . was access to history even at the price of destruction'.[13] As we observe the alt-right militias of the USA, openly carrying guns and uttering death threats against feminists, leftwingers and migrants, it is hard not to conclude that destruction, yet again, is their deepest desire. Collapse everything and start again is the modern right-wing fantasy.

Yet today's 'mob' lives in the richest country on earth; in which their rights to carry guns, protest outside abortion clinics and spout racist bullshit are unconstrained; and which is nine years into an economic recovery. Why do they want to destroy it?

If Arendt's descriptions of the dynamics of totalitarian movements hold good – and they largely do – her explanations for them do not. As a result, if Trump has triggered a crisis of progressive thought, it is in particular a crisis for the cult of Hannah Arendt. The United States of America was her last and enduring hope: the only political institution on earth that was supposed to be immune to totalitarianism, nationalism and imperialism.

Arendt's humanism was based on 'what ought to be', not on 'what is'. Human beings, she wrote, should resist totalitarianism by trying to live an active life of political engagement, and by carving out freedom to think philosophically.

But no matter how many progressive causes she espoused, hers was a worldview blighted by admiration for the reactionary German tradition in philosophy begun by Friedrich Nietzsche. Nietzsche taught the German bourgeoisie of the late nineteenth century that its fantasies of empire and *volk* were more valid than the working-class project of collaboration, equality and a human-centred society. Morality is a sham, he said, and the most honest thing to do is to pursue your own self-interest by any means necessary. There is no purpose to human existence, such as the 'good life' imagined by Aristotle, and so no set of morals or ethics can be derived from it.

Though Arendt lamented the way bourgeois morality 'collapsed almost overnight' under Nazism, her explanation for this event was, basically, that Nietzsche had been right: his 'abiding greatness' lay in demonstrating how shabby and meaningless the morality of the German bourgeoisie was, she wrote.[14]

Nietzsche would become the cult figure of neoliberalism. Once human beings are reduced to two-dimensional, selfish and competitive individuals – in a world where 'there is no such thing as society', as Margaret Thatcher once put it – the only logical response is to cast yourself as one of Nietzsche's supermen: the Alpha male, the ruthless manager, the financial shark, the pick-up artist.

Though Arendt drew different moral conclusions from those of Nietzsche, she could never see him – or the philosophical tradition he gave birth to – as the progenitor of Nazism. Indeed, she went out of her way to absolve him from responsibility for Nazism. She remained in awe of Nietzsche's pro-Nazi follower, and her one-time lover, the philosopher Martin Heidegger, until her death.

For us, understanding the philosophical through-line from Nietzsche via Hitler to the American neocons of the Iraq era and the alt-right of today is critical. Nietzsche is the all-purpose philosopher of reactionary politics. He says to the middle-class mind, dissatisfied with managerial conformity, that there is a higher form of rebellion than the one proposed by socialists, feminists and other progressives: a one-person rebellion against morality, in favour of yourself.

He tells the elite that elites are necessary, and is brutally honest that this demands a form of social apartheid in which most people perform 'forced labour'.[15] He decries state intervention, just as the modern right does, and advocates 'as little state power as possible'; he is appalled at the possibility of working people using taxation to redistribute wealth. Nietszche, instead, idolizes the 'criminal type': all the criminal lacks to be a superhero, he says, is 'the jungle, a certain freer and more dangerous form of nature' where he can demonstrate that 'all great men were criminals and that crime belongs to greatness'.[16]

Nietzsche greeted the rise of European imperialism with the words: 'A daring master race is being formed upon the broad basis of an extremely intelligent herd of the masses.'[17] What that master race needed was freedom from social norms and religious morals so that they could become 'the kind of exuberant monsters that might quit a horrible scene of murder, arson, rape and torture with the high humour and equanimity appropriate to a student prank'.[18]

Any reading of what Nietzsche actually said, in the context of the rise of the German labour movement and the birth of German imperial ambition, should leave any humanist, democrat or supporter of human rights reeling in disgust. But not Arendt.

Why does it matter? Because, if we want to trace the thread that links the barbarity of the colonial period, the widespread adoption of irrationalism among European intellectuals in the 1920s and the rise of the Nazis to the rise of the modern-day alt-right, *it is the doctrine of amoralism and biological supremacy advocated by Nietzsche.*

The Scottish philosopher Alasdair MacIntyre once wrote that there is something logical in the repeated rediscovery of Nietzsche and his superman theory. Whenever the capitalist order comes under stress and the rule of the elite is challenged, the ordinary morality that rich people profess is called into question. Repression, deviousness, lies and even murder become the order of the day. At these critical moments, the ordinary, boring bureaucrats discover that their 'morality' was just a jumble of old rules without any logical underpinning. Because of this, wrote MacIntyre, 'it is possible to predict with confidence that in the apparently quite unlikely contexts of bureaucratically managed modern societies there will periodically emerge social movements informed by just that kind of prophetic irrationalism of which Nietzsche's thought is the ancestor.'[19]

That is exactly what we are living through now – and Arendt's thought cannot explain it: because she refused to understand fascism as the elite's response to the possibility of working-class power, or to understand the essential role of irrationalism in all such reactionary movements, and because hers was a philosophy based on American immunity to totalitarian impulses, which is sadly disproved.

Arendt's optimism about post-war America ultimately stemmed from her belief that people can learn to take self-liberating actions, learn to distinguish good from bad and the ugly from the beautiful. But if you share her optimism – and I do – then you are now up against a very dangerous opposing force.

In this context, the rediscovery of Hannah Arendt and the humanism of the 1950s is not enough. We need a humanism that can resist the re-establishment of biological hierarchies and root the universality of human rights on more solid foundations than the ones currently under attack. It will need to survive contact with the new challenge of thinking machines and the new ideology of machine control known as post-humanism.

PART III

The Machines

In the midst of the self-importance of the contemporary generation there is revealed a sense of despair over being human.

Søren Kierkegaard[1]

8

Demystifying the Machine

Around the year 1600, Galileo Galilei wrote the first truly scientific book about machines. While visiting the workshops of Renaissance Italy, he was continually finding people trying to build devices that did not – and could not – work. They were all labouring under the same illusion: 'The belief and constant opinion these artificers had . . . that they are able with a small force to move and raise great weights.'[1]

People who made the pulleys, pumps and water-mills of the early seventeenth century thought machines were 'adding' something – and assumed it was energy conjured out of nowhere. Galileo's contemporary, Guidobaldo del Monte, even wrote that machines were devices for working 'in rivalry to the laws of nature'.[2]

Galileo showed them they were wrong. In forty pages of crisply illustrated maths he outlined the fundamental principle of mechanics: a machine does not amplify the force applied to it, but only transforms it. If it is powered by human labour – for example the pulley system at a wharf – it cannot do more work than the humans operating it.

In short, there is no mysterious or unnatural force operating inside a machine.

In 1776 it fell to the Scottish economist Adam Smith to establish an equally fundamental principle in economics: machines do not create value either. As the industrial economy emerged in the late eighteenth century, many people believed machines were a mysterious source of extra wealth. They believed the factory system, combined with a new technical division of labour, had somehow 'amplified' the value of outputs beyond their inputs. Smith taught them this was rubbish. In *The Wealth of Nations* he explained that human labour is the source of all value. 'It was not by gold or by silver, but by labour, that all the wealth of the world was originally purchased,' he wrote.[3]

Machines amplify the productivity of work; they allow one human being to exert force on many objects at once, and thus transform them quicker and more cheaply. But they produce no extra value: they simply transfer the value of the work and raw materials that made them into the product, said Smith. His 'labour theory of value' was the second great act of demystification achieved by scientific thought during the machine age. In economics, as in physics, there are no mysterious forces operating inside a machine.

Today we face a third outbreak of mysticism about machines. In the past three decades, the widespread deployment of information technology and the dramatic fall in production prices associated with it have promoted a new belief in the immateriality of information. We speak of cognitive capitalism, immaterial labour, virtual manufacturing and the hugely inflated accounting category of 'intangible assets'. Just as it looked to the sixteenth-century builder that ropes and pulleys were a way of 'defying nature', our laptops, tablets and smartphones, and the server farms that power them, seem to be producing intangible forms of wealth, defying conventional economics.

The existence of vast financial profits alongside stagnating GDP growth; trillion-dollar company valuations based on the ownership of intellectual property alone; 'flash crashes' in which billions of dollars can be wiped out and then restored in microseconds; the rise of digital currencies like Bitcoin – all reinforce the illusion that economic value has become detached both from machines and from labour, and can be created at will.

The myth rests on the idea that information is somehow not part of material reality. Dispelling it is important – because it has come to underpin the idea that in an information society human beings cannot be free.

A computer is a machine. The silicon chip inside it is a machine with billions of switches that do not move; a 4G network is a machine whose main components are switches and radio waves; the 'cloud' systems owned by Amazon, Alibaba and Google are also machines.

Even software is a machine and, by implication, so is a single executable line of code.

At the physical level, digital machines amplify human power over nature, just as mechanical machines do: they allow us to stack airliners in holding patterns that would be unsafe without computers; to model complex processes, to synthesize new materials, to build and 'fly' aircraft millions of times over before they are built for real. They also produce and reproduce information on a scale never possible before. This, in turn, improves human understanding of the world outside our brains, and can even equip our brains to perform better.

At the economic level, just as with previous innovations, information machines make things that were once expensive become cheap. For example, the cost of sequencing an entire genome of DNA has fallen from $100 million in 2001 to just over $1,000 today.[4] But their revolutionary potential lies in the fact that, via the same process, they can make some things that were once expensive – above all information goods – free, or cheap enough that their price barely matters.

Information technology creates goods unlike all previous goods: that can be copied infinitely at minuscule cost, used by many people at once, and used without wear and tear.

The classic example is the digital music track. Though it still has a definite production cost (the wages of the band and sound recordist, the cost of the microphones, the budget for marketing, etc.), its reproduction costs are close to zero. Meanwhile, digital technology collapses the production cost too, as electronic instruments are used, along with virtual mixing desks and precise virtual sound stages simulating conditions from the concert hall to the jazz club.

This 'zero marginal cost' effect has begun to cascade into every physical sector in which information is part of the production process, creating downward pressure on the cost of producing real goods and services. So, for example, the task of stamping metal parts with a press can now be done by robot, with the number of mistakes reduced close to its statistical minimum.[5] Or in commercial law, analytical tasks that junior lawyers once took hours to complete can now be done by a computer in seconds, leaving the remaining lawyers to sign off the results and present a human face to the client.[6]

As early as the 1990s, policymakers like the Fed's Alan Greenspan began to believe infotech was producing something that could not be captured through traditional accounting. Once you moved software into the correct column on the spreadsheet, they assumed, this would reveal that infotech would produce higher growth. But it didn't.

The OECD tried fitting the effects into something called the 'consumer surplus' – calculating how much better value customers were getting because of price competition and transparency on sites such as Amazon or eBay. But ultimately, they concluded, the biggest impact of the internet has been on 'non-market transactions', that is, activity that cannot be measured in price terms: 'These interactions and impacts contribute to individual utility and the well-being of the entire society. They are not, however, captured within the traditional measures of national accounts.'[7]

Even if you try to calculate – as the American Bureau of Labor Statistics did – the 'wages' I should be receiving if I spend time on the internet at home, the fact remains that I do not receive wages for doing so. Where people are actually paid to spend micro-globules of time on the internet, as with Amazon's Mechanical Turk labour market, guess what? The price of their labour comes under massive downward pressure, towards a dollar an hour.[8]

There is only one economic framework that can account for what is happening, and that is the labour theory of value as outlined by Adam Smith, David Ricardo and Karl Marx. They divided the quality of all commodities into 'use value' and 'exchange value', rigorously separating the usefulness of a product from the price the consumer pays. Mainstream economics says the price 'contains' the usefulness – because a price reflects what every specific user is prepared to pay at a given time. The labour theory of value says the price reflects only the amount of labour used to make the product, to feed and clothe the worker who made it, to produce the raw materials and to bring it to market.

Using this rigorous distinction between use value and exchange value, we can see very clearly what mainstream economics can't: that information technology permits the infinite expansion of use value, but tends to erode exchange value.

Used for the benefit of society, the appearance of free products over the past twenty years – open source software, open standards, Wikipedia and digital cooperatives – could massively increase the amount of human wellbeing, without expanding the market sector of the economy. This creates the possibility of a whole new journey beyond capitalism, one barely imagined by the socialists of the twentieth century, which I and others have labelled 'postcapitalism'.

Information technology, in short, makes Utopian Socialism possible: the appearance of islands of cooperative production for sharing, the massive reduction of hours worked and the expansion of human freedom and self-knowledge.

The full benefits of information technology will never show up in traditional GDP measurements or globally accepted accounting principles. Indeed the most likely outcome is that infotech depresses growth and profits as measured by traditional economics and, ultimately, the tax take. Central bankers and finance ministers, who have kept a lonely vigil over the information economy waiting for it to produce wealth, should stop trying to measure the welfare effects in monetary terms and understand that they are just use values: human benefits that are not the result of any market interaction.

However, it is not only the economists that are confused. The rise of computers has produced a new ideology of 'immaterialism' in academia over the past thirty years, which – no matter how prestigious the names associated with it – turns out to be the twenty-first-century equivalent of alchemy.

In the 1940s, the people who built the first computers changed our way of thinking about reality just as fundamentally as Galileo and Adam Smith did. They too started out trying to make better machines: Norbert Wiener designed anti-aircraft gun control systems; Claude Shannon sought to reduce the 'noise' in telephone conversations; John von Neumann worked on the atomic bomb. By the late 1940s their thinking had converged into a whole new science – information theory – which says that mathematical logic can be applied to, or discovered within, any process, from writing a symphony to building a car. As a result, all forms of communication between humans can be reduced to numbers, in uniformly sized small containers, which Shannon labelled 'bits'.

The bits can be used to measure the amount of information in an abstract way: so, for example, a Beethoven symphony is several times the size of a Jane Austen novel. In this way, all forms of communication can be studied at an abstract level, allowing universally observable laws to be discovered across very wide variations of human activity, including language and thought.

Alongside information theory there developed a specific theory of digital machines. Alan Turing, who designed the machine that cracked the German navy's Enigma code, made two proposals that have already transformed human life: that it is possible to design a physical machine to emulate specific human logical thought-processes – a computer – and that its ideal form would be the 'universal computer', a machine that can mimic all other machines and all single-purpose computers.[9]

Thanks to Turing you no longer have to carry a phone, a calculator, a digital camera and a GPS device: your smartphone carries apps that emulate them all. Even in the era where we did have to carry these separate devices we knew that one day they would all be emulated on a single silicon chip. From the moment it was technically possible to do so, computers were designed so they could run multiple programs, in separate 'windows', not just one program at a time.[10]

But Turing made a third proposal which, during the next fifty years, is set to transform human life even more fundamentally: that machines will one day be able to think.

In his 1950 paper 'Computing machinery and intelligence', Turing spelled out the possibility that, once they could process information as logically as the cleverest human being, computers would begin to out-think us.[11] He demolished the objections that they could never simulate emotions; or that they could only ever do what humans ask them to; or that they could never think 'about themselves'; or that they could never emulate the deep, subconscious 'mind'. Digital computers, Turing insisted, would do all these things within fifty years, and then begin learning independently of human teachers.

By the early 1960s information theory had begun to cascade into every other mode of thinking, including about the natural world. According to the geneticist Matthew Cobb, information theory 'put all systems on the same level, be they mechanical, organic or hybrid human–machine (as in the case of [Wiener's] anti-aircraft guns), and suggested that behaviour could be interpreted using the same principles'.[12]

Biologists began to understand viruses as collections of molecules 'programmed' to reproduce themselves by forcing other cells to 'copy and paste' them millions of times. The discovery of DNA could not have happened unless both scientists and mathematicians had begun to suggest that chromosomes contain self-replicating 'genetic information', and to treat it like a 'code' to be deciphered. Psychology, which had divided into two schools that either speculated about the unconscious mind or merely observed cause and effect in animal behaviour, gave way to cognitive science. From the mid-1950s onwards, the brain was reconceptualized as a computer and the physical neurons inside it were mapped in order to find their logical functions.

But, despite its gigantic contribution to our thinking, this 'informational turn' across science, social science and culture has generated assumptions about reality just as false as those Galileo and Smith had to deal with. Like all revolutions in science, it has reopened the debate about the relationship of mind to matter.

With the rise of the early machine economy it became common for both philosophers to use the machine as a metaphor for both society and the natural world. René Descartes wrote in 1644: 'I have described this earth and indeed the whole visible universe as if it were a machine'[13] and David Hume in 1779: 'Look round the world: contemplate the whole and every part of it: You will find it to be nothing but one great machine, subdivided into an infinite number of lesser machines . . .'[14]

The machine, and later the factory, created a new mental model through which scientists and philosophers could understand reality: the automatic process. By framing nature as a big machine these thinkers were trying to free science from religious superstition. For them, the discovery of automatic processes at work in machines, nature and human bodies was proof that God existed; that he had designed men like machines and given them the power to make machines modelled on His own power.

In 1633 Galileo was convicted of heresy for asserting that the earth moves round the sun. By the end of the seventeenth century, you could escape this fate by claiming, in effect, that reality was a machine which God had designed and then pressed the start button. Though it took a long political struggle, over the next 150 years science won this argument. It carved out an autonomous space for thinking about an ordered reality, free of superstition and random design tweaks by God.

In the history of philosophy this kind of thinking is known as 'mechanical materialism'. It is materialist because it says the world is a real, physical thing, which includes our minds; it is mechanical because it assumes the world works to a designed, logical system – and that the task of science is to discover that system. Mechanical materialism culminated in 1814, in the famous statement by the French physicist Pierre-Simon Laplace: that if you could measure the position of all objects in the universe and know the forces acting on them, then 'nothing would be uncertain'. From planets to atoms, the universe is just a giant mechanism, following predetermined and predictable laws – and that includes human history.

The final move in the victory of science over religion was to dispense with the need for God's finger on the start button altogether. This was already implicit in the work of the Dutch scientist Baruch Spinoza, who in the seventeenth century had claimed that God 'is' nature, and therefore can't be separate from it, or exist before it as its designer and initiator. But the decisive moment came with Charles Darwin's theory of evolution.

Natural selection, Darwin showed, is an automatic process – but it is not like a machine at all. It is random and without purpose. Any laws, regularities or automatic processes we see in the natural world are the product of the natural world, not of a supreme being who designed the process. By the same token, it is not our 'soul' that differentiates a human from an orangutan: if the human brain exhibits consciousness at a higher level than other primates and pre-humans, that is a product of biology plus natural selection via millions of random events. It can have nothing to do with God.

Over the course of two centuries then, scientists – by fearlessly exploring the physical world – disproved the idea that it had been designed by an outside intelligence; they disproved that the mind exists 'immaterially', beyond matter, or that consciousness is the product of a soul separate from the body.

But in the computer age, all these delusions have reappeared.

Instead of 'God', it has become common to see *information* described as the guiding intelligence of the universe – existing prior to and 'outside' nature. Scattered throughout the writings of information pioneers are the linked proposals that information is immaterial and that humans and the world around them are 'programmed' to execute an automatic process.

The basis for this return to immaterial thinking was laid by the discovery of quantum mechanics in the early twentieth century – which convinced one group of scientists that, at the most fundamental level, the physical world is created by our act of observing it. For them there are no laws of cause and effect, only uncertainty and probability – and the act of observing the sub-atomic world both changes and in a sense 'creates' it.

Though disputed – most famously by Einstein – this development in physics triggered the revival of belief among educated people that the whole world, including our physical brain, is the product of our mind. The British physicist James Jeans, who popularized quantum physics in the 1930s, wrote that the universe was 'more like a great thought than like a great machine': 'The universe shows evidence of a designing or controlling power that has something in common with our own individual minds . . . the tendency to think in the way which, for want of a better word, we described as mathematical.'[15]

Once they started writing instructions for computers in the form of code, information theorists would supply the 'better word' Jeans was looking for: software. Konrad Zuse, one of the early computer scientists, proposed in 1967 that 'the universe is being deterministically computed on some sort of giant but discrete computer'.[16]

In 1989 John Archibald Wheeler, a physicist, coined the slogan 'it from bit' – the assertion that physical things are produced by information, not the other way around. Wheeler supported the Copenhagen Interpretation of quantum mechanics, and rejected the idea of reality as a giant machine, but in an attempt to think through the problems this interpretation left unresolved, he concluded that the universe is created from information. 'All things physical are information-theoretic in origin,' he wrote: 'every physical quantity, every it, derives its ultimate significance from bits.'[17]

Gregory Chaitin, a renowned mathematician, insisted that it was only when we started writing software that we could understand the way a pre-existing intelligence had designed the universe: 'The biosphere is full of software, every cell is full of software, 3 to 4 billion year old software . . . The world was full of software even before we knew what that was.'[18]

These statements from important figures in twentieth-century science are speculative assertions, hypotheses and assumptions. They belong to a kind of thinking known as 'metaphysics' – ideas that are rational but which cannot be proved.

There is nothing wrong with metaphysical speculation. In fact, pull any bestseller on the history of science off the airport bookstand and you'll read a story of people making brilliant verifiable discoveries while at the same time uttering wild, unverifiable metaphysical claims. [19] But here's the problem with metaphysical thinking: its proposals are always conditioned by the society in which the thinker lives. If, at the dawn of the machine age, you get a theory of reality based on machines, and then, in the age of computers, you get a theory of reality based on software, it is probable that both these theories are historically conditioned, inaccurate and won't last.

Think about Chaitin's claim: it was not until we started making software that we discovered all physical reality is really composed of software. What happens when we start making something else? As the Nobel laureate Steven Weinberg wrote, if the universe looks remarkably like the computers that physicists are using in their labs, 'so might a carpenter, looking at the moon, suppose that it is made of wood'. [20]

What's important here is to understand how completely metaphysical thinking about computers has entered popular consciousness. James Gleick's bestselling book *The Information* is a bravura exposition of the new doctrine of mysterious machines. He writes: 'Every burning star, every silent nebula, every particle leaving its ghostly trace in a cloud chamber is an information processor. The universe computes its own destiny.'

As for human society, Gleick concludes, 'history is the story of information becoming aware of itself'.[21]

If you remember the ideas of Georg Hegel as he lauded the Prussian monarchy in 1818, you will see the similarity. Hegel believed in a 'world mind' becoming conscious of itself as it pushed human beings to take actions they could not understand; Gleick believes the historical actions of human beings are caused by 'information becoming aware of itself'. They are very similar theories, though separated by 200 years. The philosophical label for them is 'idealism'.

Why some of the greatest tech innovators and scientists of the mid-twentieth century embraced the primacy of mind over matter is easy to understand once you consider the problem quantum physics introduced into science. The Copenhagen Interpretation of sub-atomic physics says that, at the deepest level, reality is created by our acts of observation. It would be meaningless, said the physicist Niels Bohr, to speak of the state of two entangled particles before they were measured. If so, at this level of observation, there is no objective reality against which we can measure theories and claims about cause and effect.

Within physics, the most effective opponents of the Copenhagen Interpretation, including Einstein, focused on its alleged 'incompleteness'. What if, said Einstein, there is a deeper reality beyond this paradoxical phenomenon, which we have not yet discovered? It is far more likely, he said, that there is cause and effect at work, and that there is a reality which exists before and after we observe it, and that the Copenhagen Interpretation will one day have to be abandoned.[22]

Echoes of that debate still rage within physics, at a theoretical level most ordinary people cannot understand. However, we already use technologies that operate using quantum mechanics, like MRI scanners and ultra-thin silicon chips – and in the future, vast improvements in processing power are promised by quantum computers.

The problem is, the pioneers of quantum mechanics also made more general, philosophical claims, most notably that their discoveries invalidate cause and effect throughout all of reality; and that a reality beyond our senses, existing independently of our observations, cannot exist.

For 200 years science had assumed (a) the existence of a reality independent of our ability to observe it and (b) that whatever the latest theory says, it is only an approximation to the truth and will likely be improved on by further experiment and observation. The

Copenhagen Interpretation rejects both assumptions – since its supporters claim there can be no deeper reality than the one they describe.

For nearly 400 years philosophical idealism has functioned as a kind of parking lot for unsolved problems in science. Its main tenets were, first, the existence of an intelligence superior to humanity, which designed the universe and/or controls our destiny. Second, the idealists argued, if the world really did exist beyond our consciousness and our senses, it would be so far removed from our thinking that it doesn't really matter. In its rational form, as espoused by eighteenth-century German philosopher Immanuel Kant, idealism says we can never comprehend the 'thing in itself', only its appearance in our brain. In its extreme form idealism says, as Wheeler suggests, that acts of observation create matter.

Back in the eighteenth century the superior intelligence was labelled 'God' and our immaterial consciousness was called 'the soul'. In their place, for the same perfectly rational reasons as 250 years ago, modern followers of the Copenhagen Interpretation assert that 'the universe is computing itself' and that our perceptions create reality.

We should be wary of the fashionable allegation that 'all science is socially constructed': as we will see below, it was an accusation pioneered by the irrationalist left, which has now been hijacked by the right. However, it is obvious that all metaphysics are socially constructed.

The science historian Paul Forman documented how neatly the adoption of the Copenhagen Interpretation both fuelled and sprang from a wider hostility to science and rationality in early 1920s Germany. Defeat in the First World War caused many thinking individuals to adopt the so-called 'philosophy of life', which focused on emotion, intuition and 'fate'. They blamed rationality and science for the horrors of the war. In response, says Forman, 'one physicist after the other strode before a general academic audience to renounce the satanic doctrine of causality'.[23]

The most popular book in Germany in the early 1920s – Oswald Spengler's *The Decline of the West* – advocated overthrowing 'the tyranny of reason'. Spengler believed that science, with its emphasis on causality, should be replaced with beliefs based on destiny. He predicted that, as the West declined, we should expect a 'second religiousness' to arise, paralleling the way Christianity eclipsed philo-sophical reasoning in ancient Rome. Forman shows how explicitly some of the key figures in Copenhagen physics were influenced by Spengler's irrationalism and the wider culture from which it grew. He asserts that 'the movement to dispense with causality in physics, which sprang up so suddenly and blossomed so luxuriantly in Ger-many after 1918, was primarily an effort by German physicists to adapt the content of their science to the values of their intellectual environment'.

None of this invalidates the achievements of those who discovered quantum mechanics. But it does explain why they were so keen to turn their incomplete theory of sub-atomic physics into a complete rejection of causation across all science, with reactionary political and social consequences.

Forman, whose account was published in 1971, observed the same process under way in our own time. From the late 1960s, when sci-ence and rationality were being deployed to defend the mass killing of civilians in Vietnam, we see scientists such as Wheeler and Zuse lurch towards the idea that 'reality is a being computed', that infor-mation precedes and creates matter ('it from bit'), and that only billions of acts of observation have created the universe. This – cascaded into the popular consciousness via hundreds of articles, TED Talks and airport bookstands – has created a new form of idealism linked to computing machines.

What made the new idealism possible was the relative incoherence of those trying to give a materialist explanation of the digital world, as with Norbert Wiener, who founded the discipline of cybernetics. Wiener wrote in 1948: 'Information is information, not matter or energy. No materialism which does not admit this can survive at the present day.'[24]

If he was only saying 'information has its own laws separate from but related to the laws of physics' you could accept that: the 'laws' of a Mozart piano concerto exist separately from the physics of a piano. But he meant more: that digital information introduces a new prop-erty into the physical world unknown to physics – and that we have to understand this new thing as separate from both matter and energy.

Wiener understood that to process information you need only a tiny amount of matter and not much energy. The assumption was that the incredible power emanating from a computing machine – to solve simultaneous equations, crack codes, calculate entire company payrolls in a single day – needed a separate category in the material world.

You will often hear Wiener's famous claim repeated as if it were self-evidently true. It is not. In 1961 IBM physicist Rolf Landauer proved that information is physical and that Wiener is therefore wrong.[25] Summarizing his results he wrote: 'Information is not a disembodied abstract entity; it is always tied to a physical representation. This ties the handling of information to all the possibilities and restrictions of our real physical world, its laws of physics and its storehouse of available parts.'[26]

Specifically, he showed that information processing consumes energy and it should be possible to measure the amount of energy deleting one bit of information consumes. In their mental models of info-processing, scientists had made the act of computing a number cancel out its own cost in energy: Landauer showed it did not. In 2012 a team of scientists built a tiny model, proving 'Landauer's Rule' in practice.[27]

Information costs energy to produce and has to be represented by matter. Bits take up room in reality: they consume electricity, give off heat and have to be stored somewhere, usually, today, on a piece of silicon which is structured to retain a small electrical charge even when your computer or phone is switched off. Likewise the 'cloud' where your photo collection and music are stored is in fact acres of air-conditioned server farm space consuming – it is predicted – one-fifth of the world's energy by 2025.[28]

To say information has to be represented physically is not to deny that it has its own laws and dynamics, independent of the matter in which it is stored. So does music: a music track exists as a file on your smartphone, travels through wires, becomes sound waves inside your earphones, and stimulates electrical activity in your nervous system. And yet, from an EDM track to a Mozart aria, the meaning is created at a different level of reality: in the interaction between the physics of sound, which create patterns of tension and resolution, and our culturally conditioned brains, which give these patterns meaning.

Once we understand there is no information without physical representation, 'it from bit' becomes logically impossible – because its premise is that there was once a moment, however brief, in which information existed before physical reality.

Plus, if the universe is a giant discrete computer operating in packets and pixels, then, as Einstein observed, the whole of physics is wrong. Physics is based on the assumption that reality is smooth and continuous. When a car accelerates, or air flows around an aircraft wing, it does not do so in 'bits' – discrete packages of time and space – but smoothly. To model airflow or acceleration on a computer we break it down into bits, just as a digital photograph is merely a collection of tiny pixels. But though we see pixels if we zoom in to a selfie, that does not mean a human face is composed of pixels.

In the early twenty-first century, then, we are beset by two forms of mysticism about information machines: the belief that they create economic value out of nothing, and the belief that information exists separate from the physical world. Add to this the irrational belief that the universe is, as Jeans put it, 'a great thought' and that reality is being 'computed', and you get not only a comprehensive new form of idealism, but a strong ideological underpinning for the idea that humans are powerless, incapable of freedom and trapped within an illusory world.

The new metaphysics of science is one of the strongest underpinnings of the anti-humanism that pervades twenty-first-century ideologies. If information exists prior to the physical world, and human history is just 'software computing itself', we are back to the scenario in the movie *Jason and the Argonauts*, where every choice we make is really predetermined by the gods, moving us like pieces on a game board. There is no human freedom or agency to defend.

Why does it matter? Norbert Wiener understood why it mattered. His term for computer science was 'cybernetics': the science of control. Both humans and computers can control their outside environment. Information, said Wiener, is simply what they exchange with the outside world – orders and feedback. When humans started to build computers they at first used them to give orders to reality and receive accurate feedback from it.

But, Wiener predicted, if computers can learn – as Turing suggested they could – then one day they will give the orders to human beings. Wiener wanted to cling to a modified form of materialism because he understood that idealism about information would lead us in the direction of surrender to information machines.

Hidden within all theories about information, then, are theories about human nature and the possibility of freedom.

Luciano Floridi, the Oxford professor of information philosophy, claims that, with the emergence of networks, information technology has created a new kind of human being: the 'information organism' or 'inforg'.

Because computers can already out-think us, Floridi says, because social media platforms are anticipating our behaviours and even shaping them, they have irreversibly modified the environment in which human life takes place. If the 40,000-year history of civilization up to the present has been characterized by humans trying to control nature, we have now created something that is more in control than we are. As Floridi memorably puts it: computers 'have already begun playing as the "home" team in the infosphere with us as the "away" team'.[29]

If Floridi is right, the possibility of human freedom is already constrained: soon, computing machines will become more powerful than our brains and free will is going to become impossible. When a movement arises – as it will – wanting to place powerful artificial intelligence machines and data storage under human control, the machine owners will justifiably ask: by what right does humanity – which has already surrendered its claims to freedom, rationality, causality and the ability to act – demand control over and protection from artificial intelligence?

As they drum their fingers waiting for an answer, one faction within neuroscience will answer 'none'; so will the survivors of postmodernism, currently regrouped around the banner of 'post-humanism'; so will numerous bestselling thinkers in the popular science field. I will deal with their specific arguments below.

If the new digital idealism is right, humanism is just a form of nostalgia. If we are going to defend truth based on our sensory experience against fake news; if we are going to defend universal rights against theories of racial and gender supremacy; if we are going to replace neoliberalism with a system based on our 360-degree human needs – then for all these tasks we need to defend the concept of a human being who is capable (subject to given historical circumstances) of autonomous thought and action. Or as philosophers call it, freedom.

To do this, we must root humanism in something more solid than nostalgia. We need the things Wiener despaired of finding: a theory of reality that places digital information inside the physical world; a theory of history in which human beings, not algorithms, determine the outcome; and a theory of human nature that can refute Floridi's suggestion that we have already turned into semi-powerless 'inforgs' controlled by the machines we make.

Fortunately, such theories exist.

9
Why Do We Need a Theory of Humans?

'Man is a political animal,' wrote Aristotle in 350 BCE. Actually he didn't. Western civilization's earliest claim about human nature is properly translated as: 'man is a city animal', or more accurately still: 'humans are a species that can only achieve their true potential in a community governed by laws.'

In *Politics* Aristotle was trying both to justify and to explain the emergence of democracy in the Athenian city state. There were bigger and wealthier city states in existence in 350 BCE – in Egypt, Persia, Mesopotamia and China – but only in Athens did we see the emergence of a democracy, and a radical one at that. Granted, slaves had no rights, and women had no right to political participation, but free men – whether peasants or aristocrats – had equal rights under the law and an equal voice in the public assembly, which all were entitled to attend.

For Aristotle, it was natural for humans to live in cities, because cities help raise human life beyond merely eating, reproducing and working. A city economy, even in 350 BCE, could provide enough surplus wealth and enough free time for people to experience culture, happiness and a certain amount of freedom. By the same logic, a man who wanted to live outside the laws of a city, or to pursue money at the expense of the 'good life' of leisure and culture, was, Aristotle claimed, either subhuman or already god-like – 'either a poor sort of being, or a being higher than man'.[1]

Aristotle said each kind of being in the universe has a characteristic pattern of behaviour: when it does what it was designed for, each species fulfils its purpose, or *telos*. So his claim about men being 'community animals' is not just a description but a proposal: that to fulfil what we are designed for, we need to create communities in which we can live the good life and become fully rounded people.

Aristotle understood how critical technology would be to eradicating class distinctions. He wrote that 'if every tool could perform its own work when ordered, or by seeing what to do in advance ... master-craftsmen would have no need of assistants and masters no need of slaves'.[2] If, in other words, machines could think, learn and act independently of humans, the need for work – and the social hierarchy that goes with it – would end.

The *zöon politikon* was blatantly conditioned by the world it was created in: a city state full of 'free' slave-owners and oppressed women. It was also conditioned by Aristotle's pre-scientific outlook on reality, which tended to ask about all things – from trees to rivers to humans – 'what is its purpose?' rather than 'how does it work?' Once the slave-owning city states of ancient Greece and Rome collapsed, the *zöon politikon* was eclipsed by concepts of human nature rooted in the great monotheistic religions of the Middle East. Later, when it was rediscovered by Christian and Islamic scholars, Aristotle's political and ethical views were adapted to their religious schemas: being a good citizen meant obeying religious law; living virtuously meant abandoning the sexual and social pleasures Aristotle had identified as the good life.

For more than 2,000 years one religion after another told us that 'human nature' is immutable. There is a body and there is a soul separate from it; the soul has to be redeemed by ethical actions taken by the body. The most popular version on the planet is the one I grew up within: Christianity. It teaches that all humans are born evil (because of the original sin of Adam and Eve) but that they can be made good by obeying a set of rules and following certain rituals (baptism, communion, confession, the last rites, etc.). Once the body dies, the soul faces a binary outcome – heaven or hell for eternity – depending on the judgement of God. In case you think this last bit of the story is optional, a picture of it is painted on the wall of almost every large Christian church, including in Catholicism's HQ, the Sistine Chapel in Rome.

Islam and Judaism also believe in this body-vs-soul story (though they reject original sin). Others, like Buddhism and Hinduism, allow for the same soul to move through different bodies rather than end up forever damned or saved.

Aristotle could at least demonstrate his theory of human nature from experience: the people around him behaved politically and acted as if freedom were achievable in this life. Neither the 'born evil' part of Christianity nor, obviously, the post-death judgement process and afterlife can be proven through experience. It is in fact superstition, again conditioned by the historical circumstances of the people who evolved the monotheistic religions of the Middle East.

But if the religious view of human nature is based on superstition, and Aristotle's view is based on the brief experience of a vanished Athenian city state, what is left?

We could reject the very idea of human nature, saying 'humans are just a collection of bones, brains and DNA that tend to act in the following ways'. But once you study the way the human species acts, it is spectacularly different from all other collections of bones, brains and DNA.

For one thing, in the space of just 200 years this organism has built a carbon economy that could destroy the planet it lives on. Climate change could destroy up to 35 per cent of all species.[3] We have disrupted the ecosphere so profoundly that some scientists now propose the idea of an 'anthropocene' – a specific era in planetary history in which human beings have altered the way the earth works.[4]

On top of that, this collection of bones and brains can do something no other species can. It can build objects and machines and even design societies guided by its imagination. Sometimes the things it imagines are horrific – the instruments of medieval torture, hydrogen bombs, gas chambers – but this does not stop us wanting to know what is unique about human beings.

If you say 'there's no such thing as human nature', you are still expressing a theory of human nature: that our muscles and brains are simply programmed by DNA, modified by experience and by random variations in the electrical activity of the brain. You're saying, effectively, that the difference between zombies and humans is just a matter of degree.

In neuroscience this is known as the 'zombie challenge'. In 1983 a team of neurobiologists led by Benjamin Libet showed in a lab experiment that when faced with a snap decision, the brain activity initiating action takes place several hundred milliseconds before our brain registers a conscious decision to act.[5] Since conscious decision-making is the assumption behind the proposal that we possess 'free will', Libet's experiment gave birth to a school of neuroscience claiming that all human behaviour is determined and that free will is an illusion.

This view of human nature has become very popular in modern secular societies. The bestselling author Nassim Nicholas Taleb insists we are the 'playthings of randomness'. Yuval Noah Harari, another writer of science-based blockbusters, insists that 'to the best of our scientific understanding, determinism and randomness have divided the entire cake between them, leaving not even a crumb for freedom . . . free will exists only in the imaginary stories we humans have invented'.[6]

Little wonder then that social attitude surveys reveal many people are 'living for the moment', because they don't believe their actions can influence the future. The Pew Global Attitudes Survey, for example, claims that a clear majority of people in developing and emerging countries believe 'success in life depends on forces outside our control'. While a majority of those in mature democracies such as France, Britain and the USA tend to reject this view, in no developed country are there fewer than 40 per cent who subscribe to philosophical helplessness.

And increasingly – contrary to Harari – our 'imaginary stories' are becoming dominated by the themes of fatalism. *Game of Thrones* is just the latest in a long line of mass entertainment products in which humans are depicted as ultimately being the playthings of the gods. Substitute 'the gods' for bipolar disorder, and you have the entire premise of the long-running series *Homeland*. Substitute corruption and racial oppression and you have the subtext of the closing sequence of every series of *The Wire*. No matter what the criminalized black men of Baltimore do they cannot escape their fate, nor can Carrie Mathison escape her compulsion to save American imperialism while destroying herself, nor are the rapes, murders and intrigues of *Game of Thrones* anything other than the results of fate.

The logical flipside to our growing belief in fate over freedom is the rising obsession with gambling: luck, not purposeful activity, is the only way to cheat fate. Under the influence of pseudo-science and the worship of market forces, fatalism has become the twenty-first century's folk religion, and the online casino its cathedral.

Let's list the implications of all this.

First, if we're programmed by the reality around us, our capacity for free will is meagre. If so, all the major human-centric religions of the world go up in smoke. For it was not just Aristotle who believed in free will but the founders of Judaism, Christianity and Islam: through our ethical choices we can be redeemed, say the monotheisms of the Middle East.

The second problem concerns our ability to achieve our human potential: to be happy in both our work and our free time. For Aristotle human beings were perfectible only in a city. For the Christian theologian St Augustine they were perfectible only in the 'City of God' – that is, beyond their existence in the physical world but still as part of a community. For people who believe human beings are completely shaped by their DNA plus their surroundings there can be no perfectibility – or if there is, there has to be a force acting on us from outside to make it happen. Our own choices do not matter.

This in turn creates the ethical vacuum that currently surrounds us. For all versions of the DNA/neurons/fate theory of human nature, morals are optional. You can borrow a specific moral rules-set, copied and pasted into your life from an old religion, but as you break it, vigorously and often, you do not expect your behaviour to have any ultimate impact on your prospects, here or in the afterlife. Not for nothing did Cersei Lannister, the amoral manipulator of other humans, become the most compelling character in *Game of Thrones*. Many in the audience looked at Cersei and thought: 'she's evil, but she has fun and she survives'.

Fortunately, those of us who want to reject the folk religion of fatalism also have strong scientific and philosophical traditions to draw on. Harari repeatedly claims that neuroscience supports the idea that free will is impossible. There is, however, an entire neuroscientific literature that refutes this claim – a literature rich in insights into the uniqueness of human consciousness compared even to that of higher primates.

Libet's 1983 experiment, which has since been confirmed by observing single neurons in the brain, found a rise in brain activity around 500 milliseconds before we become conscious of what seems to us like a free, voluntary decision to act. This led him to conclude that, if we have any freedom at all, it is in the 150 milliseconds during which we can override our response to such brain activity. This makes our actions, ultimately, the product of our biology plus our environment. Furthermore, other neuroscientists found reliable evidence that once we've acted we then rationalize the action as a decision: our biology, in other words, not only robs us of free will, but actively creates the illusion that we have it.[7]

These conclusions have been criticized in two ways: first, by psychologists who insist that this pre-programmed 'readiness potential' reflects a capacity to act that is the result of numerous previous decisions and experiences, which we store and draw from as external events require us to make decisions.

Second, within neuroscience itself, more recent experiments have shown that the build-up of brain activity that happens before making a decision may simply be a random surge of activity that is common to the nervous systems not only of primates but even crayfish – who of course cannot make conscious decisions at all. Aaron Schurger, a Lausanne-based neuroscientist, concluded: 'We may have been wholly wrong in our assumptions about the nature of the brain activity that precedes voluntary movement, for 50 years measuring, analyzing, and mapping what may turn out to be a reliable accident.'[8]

As the neuro-philosopher Andrea Lavazza points out, the random brain activity that happens before we make a conscious decision is the result both of our brain's biology and our past experiences, which include implicit knowledge of what happens if we make a certain decision and act on it. The latest experiments not only bring the neuroscientific study of decision-making back into the realm of the psychological and the social. They accord with how we intuitively understand what we are doing. 'When one forms an intention to act,' writes Schurger, 'one is significantly disposed to act but not yet fully committed. The commitment comes when one finally decides to act ... with the decision to act being a threshold crossing neural event that is preceded by a neural tendency toward this event.'

In summary, the neuroscientific evidence to support the absence of free will is not conclusive. Those using it to bolster the philosophical claim that humans have no capacity for freedom do so because they are predisposed to that particular view of human nature.

Is there any theory of human nature that allows for the possibility that we will perfect ourselves in this world, using our amazing brains to imagine solutions to the problems of hunger, desire, unhappiness? And to do it ourselves, without the intervention of God or a giant computer programmed before the start of time?

To construct one, you would have to start with a list of unique biological facts about human beings. We learn and don't stop learning. At around the age of two our brains stop simply reacting to their environment and begin to develop a consciousness of a 'self' and others, which can be expressed through language.[9]

We teach each other to reason – to make conscious, reversible choices between two or more actions. This ability to do 'operational logic' develops through trial and error between the ages of five and seven and later becomes possible to do in our heads.[10] We make things – but in a different way from all other living things: we can imagine the thing to be made in advance and create the tools to make it.

As with chimps and baboons, our biologically available advantages develop properly only if we live in ordered, hierarchical groups. But unlike other primates, human beings can consciously change the structure of the hierarchical groups we live in, and even reject hierarchy entirely.

Finally, humans have an advanced capability to communicate through language which, as far as we know, no other species matches. Our language is the product of consciousness, imagination and sociability. A robin's call will change depending on whether it is in the city or the countryside. But a robin cannot decide to change its call at will. Humans can: our language can not only describe the world around us but imagine how the world might be different. From early in childhood we can say not only: 'Mum, look, there's a bird!' but 'Mum, I am a bird and I am flying over the city and I can see everything below me.'

These are the biological attributes summarized in the label *Homo sapiens.* We shared most of these attributes with other human types who interacted with our ancestors before the last Ice Age. Both we and the earlier humans made stone tools. However, it is likely that our better ability to imagine things, combined with our ability to communicate via language, is what allowed *Homo sapiens* to begin creating cultural objects around 40,000 years ago.[11]

This account of 'what makes us human' is based on the best science available. If true, it means that all the societies, cultures, languages, imaginative stories and ethical systems ever created by *Homo sapiens* need to be factored into the biological definition of our species. It means we are biologically programmed to be social, to learn, to produce a history out of the myriad choices made by the billions of humans who have ever lived.

Put another way, even if neuroscientists have proved there is an unconscious impulse at work milliseconds before we make a choice – an attribute we share with less conscious animals like beavers and chimps – there is still a question to answer: why did humans build the Parthenon and chimps not?

If it is in our DNA to tell stories about mythical gods and to carve bone objects representing them, then these stories and objects must form part of the definition of human nature. But the stories and the objects change over time – and this brings us to a basic observation: *Homo sapiens* is a species which is biologically fairly constant but socially changeable.

Human nature changes during history, according to the world we live in: the technologies, the class structures, the cultures, the norms of behaviour. Of course there can then be reciprocal changes in brain structure and function, and physical improvements as our diet and health improves, but ultimately we're still biologically similar to people who lived 50,000 years ago.

And yet we are not mere products of our surroundings: all humans have the capacity to think 'beyond' their surroundings. The capacity to imagine what's not there is constant, and is indeed a very strong impulse when our environment fails to deliver basic necessities such as food, safety or security.

If you accept that this capacity to imagine better social arrangements and to create them is not the product of a 'soul' but a function of a physical organ called the brain, you have to at least take seriously the first philosopher who properly explained it: Karl Marx.

Marx lived in a Christian-dominated society. The vast majority of those around him believed humans had 'souls' separate from their bodies, that they were born evil, and that moral behaviour had to be coerced into them by priests. Marx arrived at Berlin University in 1836, just five years after the death of Hegel. Hegel, as we've seen, taught his students that history was simply the unfolding of an idea in the mind of God and that God's mind had run out of new ideas once a liberal Prussian monarchy was running Central Europe.

But by the time Marx turned up in Berlin, Hegel's younger followers were already ripping the great man's doctrines apart. One had written an alternative life of Christ, claiming Jesus was just an ordinary man. Another argued that deities were merely inventions of the human brain, religions were just projections of our fears and failings – and that maybe Jesus was therefore also an invention.

Marx came to an even more radical conclusion: that history is not the unfolding of the world mind, or God's will, but the unfolding of the biological potential within human beings to change the world around them. Human nature changes as we transform the world around us. We can change human nature by changing society. And it is this biological attribute that gives us what Aristotle called a *telos*, or purpose. Marx said the biologically given purpose of human beings is to set themselves free, using technology to change both their environment and themselves.

Marx is famous for many things: a manifesto predicting the revolutions of 1848, written two months before they started; a 3,000-page book about the workings of capitalism; and founding an international workers' party. But if I could rescue only one of his achievements it would have to be his first: a clear definition of human nature that is compatible with our biology, our history, with technological change and with current advances in neuroscience.

Humans, said Marx, differ from animals because they can imagine changes in their own environment, express them through language, and execute them through work. Given Marx's obsession with work and with workers, you can be certain that if he'd wanted to define human nature simply as 'the ability to work' – as Benjamin Franklin did with the Latin term *Homo faber* – he would have. Instead he defined it as 'species being'.

Marx said: every time we imagine a change we are going to make to our environment, we confirm in our own minds that we, and all other humans, have a certain amount of freedom to shape our environment. When we make that change, we do so through social activity: whether we're working in a windmill, a factory, a military airbase or in our bedrooms via a network, our tools and workplaces are typically social. So when we work, we work on behalf of all other humans. 'Man is a species-being,' Marx wrote, 'because he treats himself as a universal and therefore a free being.' Combining these insights he insisted: 'Free, conscious activity is man's species-character.'[12]

This is Marx of the early 1840s, before he plunged into detailed writings about economics, before he became an active revolutionary, and before he was completely engaged in working-class politics. For the early Marx, communism meant simply the realization of human nature. In 1844 he defined it as: 'the complete return of man to himself as a social (i.e., human) being – a return accomplished consciously and embracing the entire wealth of previous development'. Communism, Marx wrote, 'equals humanism'. What stood in its way was private property and the power relations that go with it.[13]

The Marx who wrote this was unknown to the intellectuals who formed the first socialist parties in the 1880s, to the workers who staged the Russian Revolution, or to Lenin, Stalin and Mao. When the humanist essays of the young Marx were discovered in the 1930s, they were politely ignored and labelled the 'early writings' by the official communist world. In China it was not allowed even to openly study them until the late 1970s.

This freedom-centred definition of communism didn't exactly fit with the world of the Five Year Plan and the mass incarceration and murder of political opponents. Nor did Marx's early philosophical works say much about the doctrines of inevitability that had become associated with official Marxist philosophy. But the first person to translate them into English grasped their power. 'Marxism,' wrote the self-taught American revolutionary Raya Dunayevskaya, 'is radical humanism.'[14]

*

We will come back to Marx and give many of his other ideas a theoretical kicking. But it should be obvious now why this Marx, and the radical, left-wing humanism he inspired after the Second World War, is becoming relevant again.

Liberalism is not only under attack: it looks increasingly incapable of defending itself. The core proposal of liberalism is that there is a single, legal 'self', which has rights and responsibilities and the capacity for autonomous thought and action. Liberalism's idea of 'free will' was always centred on the proposal that human beings have the power to make moral judgements and take responsibility for them, and that a market-based society offered the highest form of freedom. Ours is the freedom to choose not only between good and evil but between Nike and Adidas. For Hegel, our very ability to exercise free will was dependent on the ability to own private property: to embody moral decisions through buying and selling, and to profit from our morality by owning stuff.

The proponents of free-market economics, above all Friedrich Hayek in *The Constitution of Liberty* (1960), reframed the issue: for Hayek freedom is about having the smallest possible state and avoiding the attempt to apply rationality to social outcomes. The best society, Hayek argued, is one that emerges spontaneously. The spontaneous outcome of millions of acts of free will is a better way of achieving freedom than trying, for example, to suppress inequality through a welfare state, attack poverty with wage and price controls, empower workers with trade unions and so on.

Ironically, once they were adopted as a justification for the neoliberal system, Hayek's ideas gutted classical liberalism of all meaning. Once people began to believe that a 'spontaneously emergent order' – the market – was more just, more intelligent, more humane than one rationally designed by democratically elected government under the pressure of demands for social justice, they began to lose interest in free will and in moral judgements, and ultimately in democracy itself.

In the space of three decades, the ideological propagation of Hayek's doctrines has produced a mass conversion to fatalism: the market knows best, all politicians have to be its servants, attempts to improve human society by design lead to gulags and concentration camps – this is the new common sense. As I suggested in the introduction, the danger is that submission to the logic of the market becomes a gateway for submission to the logic of the machine. Both are created by humans; surrender to their control can be justified on the same basis.

Hayek was devoted to the rule of law. He believed it was the underpinning of all the freedoms capitalism provided. Yet in the space of three decades, the coercive introduction of market forces into everyday life in Hayek's name has spectacularly eroded the rule of law. The very politicians suspending constitutions, attacking the press as 'enemies of the people', and assembling kleptocratic empires do so in the name of a market-based concept of freedom.

Liberalism – in the shape of globalist political centrism – has become a form of nostalgia and denial. 'Progress is real', shout last-ditch defenders like Steven Pinker: look how many people went from one dollar a day to two dollars a day in the past half-century; look how few people actually die in the wars we've unleashed on the world; look how shiny the Fourth Industrial Revolution will be when it actually arrives. But the rise of right-wing authoritarianism and irrationalism are evidence that the arguments of liberalism's defenders are failing to bite.

The connection between support for politicians like Trump and a fatalistic attitude to human nature is well evidenced. Meanwhile, the fatalism promoted by bestselling authors such as Taleb and Harari will, if unchallenged, leave us disarmed against the ongoing power grabs of tech monopolies and surveillance states. The same fatalism pervades the neo-Confucianism of the Chinese state as it prepares to install a social control system, linking all behavioural data to a social insurance 'score', which can bestow or deny access to jobs, education and travel rights.

In the early twenty-first century the attacks on human choice and freedom are merging into a single project: technologically empowered anti-humanism.

Liberalism, which claimed the human capacity for free will was eternal, has no defence against the actuality of its erosion in given historical conditions. And because it spent the past thirty years telling us no other system was possible, it is devoid of any political strategy except defending the status quo.

Only if you believe, as Marx did, that freedom is going to be a social and historical construction – not an innate quality – can you begin to see a way of regrounding society on human values, not machine values. But a theory of human beings takes us only to the threshold of the main problem: the challenge of machines that can emulate us.

10

The Thinking Machine

When it came out in 1976, the video game Breakout changed the world.[1] I remember the thrill of playing it as a teenager in the gaming arcades: it may only have involved hitting a ball with a paddle against some tiles, but it instantly made all mechanical pinball machines as uncool as your dad's cardigan.

In 2013 researchers at the artificial intelligence group DeepMind Technologies wrote a computer program that could learn to play Breakout. With no prior knowledge of the rules, no sight of the computer code, and using only what it saw on the pixellated video screen, the computer quickly learned how to beat the typical score of an expert human player.[2] It was given a goal – optimize the score – and it succeeded.

That's a long way for technology to travel in thirty-seven years. It's hard to explain to people from the digital era what it felt like the first time you interacted with a screen. I remember it feeling like an instant extension of reality. I also remember it changing – again instantly – the sociology of a pinball arcade. Up to then the top dogs had been muscular, rough kids good at 'tilting' the machines through physical strength (and in defiance of the rules). You cannot tilt at Breakout. Soon the top dogs were the silent, studious nerds.

But Breakout was not even yet running on a true digital computer: the men who designed the Atari machine used a 12-inch electronic circuit board full of wires and transistors. Their names were Steve Wozniak and Steve Jobs, and the next thing they built was a personal computer, which they named the Apple I.

By the time they built the Apple II computer Wozniak had worked out how to emulate Breakout as software. When DeepMind taught their computer to play Breakout, they did so by emulating a much more complex set of electrical circuits: the human brain. Since the 1950s, scientists had been trying to construct artificial neural networks (ANNs): circuits of computer processors that mirror the multilayered relationships between the neurons in our brains.

In an ANN there is always an input layer and an output layer, which in Breakout would be a processor emulating what my eye sees and another one emulating the braincells that control my hand. Between these two layers there are 'hidden layers' emulating the way my brain works at different levels of abstraction.

So, for example, when I play Breakout my brain is simultaneously asking: where is the ball? Where is the paddle? Am I winning? How did I win last time? Or it's reminding me: hey, when there are few tiles left the ball speeds up; or hey, you have only one 'life' left so be careful. The brilliance of human brains is that they can function at many different levels of abstraction at once. By 'abstraction' we just mean 'making a certain kind of sense' out of information. If you categorize each kind of thought-pattern as a 'hidden layer', which can randomly talk to any other layer, you have a working model of a brain playing a computer game.

When scientists tried to build computers to make these decisions using complex logical reasoning, however, they failed. Normal computer programs, like the BASIC on which my generation learned coding, work through algorithms that ask 'IF, THEN': if the ball is travelling left at medium speed, then move the paddle left. But 'logical AI' could develop only slowly. So for decades the quest for artificial intelligence progressed in the form of task-specific programs: software to recognize handwritten zipcodes, for example, or to play chess. The computer had to be trained, taught the ideal outcome. Its learning was 'supervised'.

In 2006 a combination of new thinking and greater access to raw computing power kickstarted new research into an alternative method, one which had hit a dead end in the 1980s. Instead of trying to emulate logic using maths, scientists tried to emulate the physicality of the brain using silicon chips. With a lot of data, a lot of processing power and a lot of random connections between layers, logic doesn't have to take all of the strain.[3]

You could, for example, load a file containing every possible relationship between paddle and ball, and ball speed, each labelled 'good' and 'bad'. The computer searches through this training data randomly, learning in a way that fits its logic capabilities. Essentially, as one eminent AI boss put it to me, 'it is playing Snap!'

Instead of the classic decision tree – shall I go left or right? – a deep-learning system contains a wonderfully named solution called 'random forests'. Here, the computer is encouraged to learn by making mistakes, rather than continually searching for the right answer.[4] If you've ever sat in a foreign language class you'll understand the principle: hearing thirty people make thirty different mistakes with the same sentence is a much better way to learn than being told the right answer in one-to-one tuition. As a result, each layer of an ANN trains itself to recognize reality at a different level: for example, pixel, ball, speed, game, rules, victory.

When Wozniak built the Breakout circuit board, Atari offered him a bonus if he could use fewer than fifty transistor units. Though he achieved the target, the machine turned out to be easier to manufacture if they used a hundred. The breakthrough in neural network design came when scientists realized they could operate the same industrial principle: with more processing power and more data storage you can throw layer after layer of electronic brainpower into the mix.

In the past ten years, then, the quest for artificial intelligence that began with Alan Turing has accelerated. 'Big data' means more than simply the ability to process and store lots of information. Once we have a machine that can learn – unaided – by crunching through data, the bigger the pile of data the more useful it can become. The ideal size of a data store for, say, an artificially intelligent chess player is every game ever played and every possible game that could be played: the data store would then contain a solution to every possible situation.

The milestone came in 2016 when DeepMind, by now acquired by Google, designed a program that beat the highest ranked Go player in the world. Computers had long ago 'solved' draughts; and IBM's Deep Blue computer had beaten chess grandmaster Garry Kasparov in 1996. But the game of Go is massively more complicated than chess: there are more potential combinations of stones on the board than there are atoms in the universe (a fact that it was only possible to calculate in 2016).

DeepMind's AlphaGo programme beat Lee Sedol, the world's top player, 4–1 in a dramatic live showdown in Seoul. After attacking aggressively and losing in game one, Lee was operating a cagey strategy in the second game when, at move thirty-seven, the computer hit him with a move no human player would have made. Reviewing the operation later, DeepMind's programmers realized the computer had asked itself 'what is the least likely good move a human would make?' and found one that had been made only once in 10,000 games.[5]

Expert onlookers judged the move 'beautiful'. Lee was so shocked he had to get up and leave the room. A clearer example of Luciano Floridi's metaphor about 'inforgs' – that humans are the 'away team' in the digital world – is hard to imagine. In despair, 'many people drank alcohol', reported the Go correspondent for a Korean daily newspaper. 'Koreans are afraid that AI will destroy human history and human culture.'[6] Were they right?

What is certain is that, with unsupervised machine learning, humanity has created a tool unlike any other. From stone tools to attack drones, we have always made tools we can control, and whose workings we understood – even if, like a drone, they operate automatically. Artificial intelligence, even in the 'weak' forms being deployed nowadays, is different. It will, say those currently developing it, have a tendency to escape from human control. And parts of it are technically not observable: as it learns, humans lose sight of how it works.

Fear and loathing of machines, a major sub-theme of modernity, began a lot earlier than *Terminator*. In the first modern novel, Miguel Cervantes's *Don Quixote*, written three years after Galileo wrote his thesis on mechanics, a Spanish knight attacks a windmill with his lance. Windmills, by the early 1600s, were not a new technology, but they were part of a mixed economy of early industrial production, milling grain, tobacco and spices, and sawing wood. Ranged in a line across the plain, Cervantes's windmills would have been a local concentration of technological power, skills and industrial knowhow. Don Quixote attacks them because he cannot understand what they are. If he did, he would know that the new commercial economy the windmills represented was a threat to his entire value system and culture.

Throughout the back half of the twentieth century popular culture played with the idea of the threat of artificial intelligence. Philip K. Dick's *Do Androids Dream of Electric Sheep?* – which became the movie *Blade Runner* (1982) – envisioned a time when androids are distinguishable from humans only by their incapacity for empathy with each other, and for animals. This, Dick imagined, was the one thing you couldn't program a computer for.

In the novel, the agents tasked with killing escaped androids use a fictional test modelled on Alan Turing's test for artificial intelligence: the Voigt-Kampff test. It was based on the assumption that humans understand the impact of all events on their species, while androids do not. 'As long as some creature experienced joy, then the condition for all other creatures included a fragment of joy,' muses Dick's hero. In short, Voigt-Kampff is a test for what Marx called 'species being'.[7]

But Turing himself had specifically ruled out this test. For Turing any human quality, including emotion and self-consciousness, could be emulated by a machine. The fact that, in *Blade Runner*, the Terrell Corporation does not program its androids to show empathy is a choice: a safety switch.

If AlphaGo's achievement in beating Lee Sedol really is a world-changing moment, let's consider the human choices that surrounded it. First, Sedol chose to play against AlphaGo. He could have chosen not to play, thus depriving AlphaGo of the experience of learning from the best human being.

Second, in order to stage the match on a real board, a human player acted as AlphaGo's intermediary. The human player, instead of merely obeying the computer, could have used its moves as suggestions only. That, too, is a human choice.

Third, DeepMind Technologies could have chosen to handicap their computer: feeding it limited information, limiting its ability to learn.

Fourth, Lee Sedol could have asked for his own copy of AlphaGo and programmed it with his own specific playing style: by combining the best computer with the best human brain available, he might have hoped to defeat a side containing the best computer only.

Fifth, the Go-playing community could have used Sedol's defeat as a sacrifice from which to learn new playing styles from the computer itself. This, in fact, is what's now happening, with players adopting advanced strategies modelled on the machine's strategy. At the top level, some claim, the AlphaGo–Sedol match has changed the dynamics of a game that is thousands of years old.

However, in the meantime DeepMind had redesigned the program: instead of sifting through hundreds of thousands of previous games, the new iteration, AlphaGo Zero, learned by playing against itself. Within three days it beat the machine that had beaten Lee Sedol. Within forty days it achieved the highest Go skill ranking in history. As its designers explained, the machine 'is no longer constrained by the limits of human knowledge. Instead, it is able to learn *tabula rasa* from the strongest player in the world: AlphaGo itself.'[8]

These choices – to refuse engagement, to mediate artificial intelligence by human decision-making, to slow down its development, or to accelerate human learning from it – constitute the logical responses to the development of AI. But in the meantime it is going to go on improving, if we allow it to, independently of our choices. Failing all else, we can attack it with our lances, as Quixote did in dumb incomprehension at the windmill. But that would be a bad idea, and just as futile.

In its first 200 years, industrial capitalism enabled human productivity to take off. In the past fifty years, a combination of computing power, globalization and rising educational levels have allowed the benefits of rising productivity to cascade over to the underdeveloped world and the global south. But this could be just the prelude to a decisive human take-off which propels us towards economic abundance.

If we can move artificial intelligence beyond its current showcase deployments, and use it to design and run the systems we need to survive on this planet – from smart energy grids to smart cities to synthetic medicines – then Aristotle's daydream becomes realizable. Machines that know their tasks and can do them without human guidance could begin to obliterate class divisions, hierarchies, poverty, oppression and inequality.

But here there arises a mismatch between what regulators think they need, what the engineers developing the AI think they need, and what society actually needs. There are, even today, no clearly agreed and implemented global safety standards for AI. There are numerous strands of academic and professional work under way to create basic safety rules – for example, the IEEE 7000 standards on AI safety, transparency and so on. But nobody is obliged to follow them.[9]

Companies like DeepMind have 'ethics committees' – but their work is non-transparent and does not appear anyway to be guided by clear ethical statements. In any case, the model of the 'ethics committee' does not suit artificial intelligence in the same way it does medical experiments: the world of pharma and biotech is, for now, dominated by closed, goal-oriented projects, such as finding a cure for cancer or a treatment for diabetes. Artificial intelligence is a general technology answering open questions, indeed questions that humans may not even be able to formulate.

We urgently need clear safety codes and a code of ethics which places all artificial intelligence being developed under meaningful, observable and irreversible human control. But such is the power of the new technology that this cannot be done at the level of individual teams, firms or – unfortunately – countries. If we develop AI under ethical control in country A, while country B is doing so without ethical control, we simply hand country B the ability to steal, destroy or otherwise sabotage the ethical form of AI. For this reason ethical use of AI is either going to be a mandate at a global level, or not at all.

Capitalism – which has regarded the ethical use of machines as 'nice to have' for the past 250 years – now faces a strategic problem: it cannot, even to its own shabby standards of prudence and safety, deploy this epoch-making technology without erecting new controls at the social level. Yet it has spent decades trying to expunge morals and ethics from economic decision-making.

Artificial intelligence, machine learning and robotics bring humanity face to face with issues we assumed could be outsourced to religion, philosophy or the self-help manual, or solved functionally by boards of experts. Such is the potential power of the thinking machine that we cannot take the next step forward without deciding who we are, and what values we want our machine intelligence to express.

To understand why, let's imagine a machine like DeepMind applied to a real-world economic and social challenge.

For around 2,000 years we have cultivated apples using a technology called the orchard. Right now we're producing around 84 million tonnes of apples per year, and orchards occupy 5 million hectares of the earth's surface.[10]

An orchard consists of trees planted to create a microclimate, with one kind of apple grafted onto the root stock of a different kind, and the apples monitored by the human eye and picked by the human hand. In the twentieth century we improved this technology by using industrial pesticides and fertilizers. In the computer age we've added the barcode and automated the back offices of fruit farms. But the basic problem remains: to pick the apple you need to know that it's ripe, and to be able to detach it gently from the tree. For this reason, the technology of the orchard hasn't changed fundamentally since it was invented, and tens of thousands of people are employed doing the back-breaking and chemically hazardous work of industrial fruit farming.

In 2017 the first prototypes of an automated apple-picking machine were deployed. The machine senses the size and ripeness of the apple and – when it's ripe – sucks it down a vacuum tube, which is guided towards the apple via robotic arms and yet more sensors. Few people would regret the replacement of back-breaking manual labour by a machine, if it can be made to work. But the apple-picking robot is a great example of how crude most robotization projects up to now have been. It simply automates a cumbersome human process.

Once we have developed AI that can consistently out-think human beings, as AlphaGo did with Lee Sedol, the solution will be to show the computer an apple and ask: what is the best way to produce 84 million tonnes of these?

The computer might specify artificial sunlight, or nourishing the fruit tree's roots with gas and liquid sprays rather than soil. It might come up with a way to manufacture apples from other compounds. It might ask: why do you need so many apples?, since the combination of sweetness and bitterness needed to make cider, for example, might be achieved by chemical synthesis. But in every case, the computer's answer would depend on how a human being defined the word 'best'.

We might ask: what is the best way to grow apples while preserving the natural environment of the valleys and fields they now grow in? Or: what's the best way that minimizes the use of fertilizer and pesticide? What's the most carbon efficient way of growing them? How could we do it with the minimum amount of work? After more than 200 years of the factory system we know how a factory is supposed to be regulated, what the international standards are, what best practice is. But we know none of these things about autonomous intelligent machines.

So the crucial question is: what do we mean by 'best'? But this is where the problems start. Because our society is already swamped by problems of choice and design caused by competing definitions of 'best'. Though some of them look like choices around cost and quality, all are at root ethical choices.

When you walk into the supermarket to buy apples you are implicitly working through a set of questions. Which are the cheapest? Which are the best quality? Which are organic? Which travelled the fewest air miles? Which ones do I usually buy? Which are the ones my mother used to give me? Which ones are the easiest to reach as I rush through the store on the way home? If the cheapest apples were picked by super-exploited farm labourers living in a tin shed in Spain, do I care? Even if you don't, your decision reflects a particular ethical mindset.

The problem is, though everyone has a rough idea what buying an ethically produced t-shirt entails, and even what kind of decisions a medical ethics committee might take, when it comes to building an autonomous intelligent machine, you can't just buy a set of ethics off the shelf. To frame the ethical development of AI you would need a set of ethics that conforms more closely to a complete moral philosophy.

But few corporations employ moral philosophers. Studying the subject at university is not exactly route one to a high-flying career – unless you have aspirations to become a bishop. But the mere possibility that we could create relatively autonomous, or 'strong' AIs within this century means we are forced to confront the moral implications at a systemic level.

When it comes to ethical systems, in our everyday lives we experience them in roughly four flavours.

The one best known to the shopper in the supermarket is called Utilitarianism: the ethical choice is the one that leads to the greatest happiness for the largest number of people, while doing the least harm. Utilitarianism was popularized by the British liberal John Stuart Mill in the 1860s and became embedded in the ideology of Anglo–US capitalism, both via philosophy departments and as a form of common sense.

Using this outcomes-based ethical system, you might ask the artificial intelligence to measure the number of air miles used to fly apples from Chile to your supermarket, against the poverty it would create in Chile if the fruit industry went bust.

However, the shopper may well have heard of a second system based on 'social justice', and may even have heard of an American philosopher associated with the term, John Rawls. Rawls said that our ethical systems have to be based on an eternal and rational set of expectations common to us all: maximum freedom and maximum fairness. Rather than let each person make single judgements about what achieves the 'greatest happiness for the greatest number', society should guarantee everyone a set of basic social and economic rights. In addition, if there are social and economic inequalities, the justification for them is that they benefit the poorest most. This is not an ethical system as in 'a way to live your life', but a social contract designed to create an ethical society.

If you program an artificial intelligence according to this rules-set, it can override the Utilitarian outcome. Indeed it will override a whole number of outcomes, vetoing numerous potential innovations on the grounds of their fairness to people alive today. And its assumptions are that both the human beings it is serving and the society they live in remain – as for Utilitarianism – self-interested individuals trying to negotiate their way to an optimum form of a market economy. Nothing in social justice ethics mandates the computer to eradicate inequality or scarcity.

A third broad ethical system in use today is the one associated with Friedrich Nietzsche. This says all ethical systems are a sham, that humans have little or no free will and that they should pursue their own happiness, if necessary, at the expense of others and by breaking every given moral code. Nietzsche tells a self-selected group of 'higher types': live for yourself and use others for your own ends. Listing the attributes of a great man, he says: 'he wants no "sympathetic" heart, but servants, tools; in his intercourse with men, he is always intent on making something out of them'.[11] As practised by its adherents in Silicon Valley or the internet trollosphere, it could be summed up as 'fuck you ethics'.

Taking Nietzsche's ethical code into the supermarket, a shopper might be inclined simply to buy the sweetest apples and let the environment and the workers die of pesticide exposure. Or, remembering Nietzsche's attitude to crime, they might steal the apples or – their will was feeling especially triumphant – shoot the cashier in the face for fun.

Finally, there is the ethical system derived from Aristotle, one concerned with virtue. According to Aristotle, all actions are judged against whether they contribute to human beings fulfilling their potential, not just individually but in a way that enables them to live the 'good life' inside an orderly political community. Virtuous acts don't only create good social consequences: to be virtuous, they have to improve the person doing them and lead the entire community towards a life of dignity, education and enjoyment. Virtue ethics, therefore, assume there is a community with the goal of living the good life. To be useful for programming artificial intelligence, that 'community' would have to be the whole human race. It is fair to say that apart from among Catholics, whose medieval theologians borrowed the idea of virtues from Aristotle, such virtue ethics are not widely in use today. And where they are consciously utilized, the focus is on individual behaviour, not the wider social outcome.

Into these four ethical systems – happiness, social justice, 'fuck you' or virtue – we could place almost any specific set of instructions for human behaviour ever invented. The question is: can any of them readily be applied to the global governance of artificial intelligence? In fact, do any of them even survive contact with it?

What's striking about the first three is how, despite their long history, they've become closely embedded in the ideologies we use to live our lives under the neoliberal system. Neither utility, social justice nor Nietzschean will to power are concerned with a project for the destiny of the human race.

It would be good if more people were happy, say the Utilitarians, but if a lot of them remain poor, stressed, mentally ill and insecure, that can still be a result of optimal ethical choices. Social justice, say the centrist politicians who idolize Rawls, is a matter of how we structure capitalism to do the least damage: there is no imperative to remove inequality, only to mitigate its impacts. As for the modern followers of Nietzsche, theirs is the philosophy of selfishness resident in the yachting clubs of the super-rich.

Only virtue ethics makes a claim based on the destiny of human beings, and judges ethical choices against a final goal for humanity. For this reason, in order to make the idea more acceptable to free-market capitalism, modern advocates of virtue ethics have reframed it as a project for specific communities. The 'communitarian movement' which emerged in the USA in the 1980s, as a response to the breakdown and atomization of communities under free-market economics, was one expression of this. To a world suddenly terrified and revolted by the idea of a 'common good', they said, the 'common good' can effectively be reinterpreted as 'what's acceptable to relatively conservative people in my town'.

However, there is a strong case for saying that virtue ethics is the only ethics fit for the task of imposing collective human control on thinking machines. We might still use a mixture of utilitarian calculation and fairness ethics to solve specific problems, but if we are looking for a set of values to form an overarching pathway to technological abundance, we have to choose between Aristotle and Nietzsche: between the good life for all or 'fuck you!'

You may not like the sound of any of these systems. You may instead live your own life to a kind of folk-philosophy roughly based on what you learned from an ancient religion, triangulated against what's acceptable to your friends. If so, you are simply relying on an incoherent mixture of ideas. And while that is fine for an individual, it is not adequate for an entire species suddenly confronted with machines that could soon out-think us.

How might the above four ethical systems be used to formulate the question we want to ask the artificial neural network: what's the best way to produce 84 million tonnes of apples per year?

Outcome-based ethics should be easy to code into artificial intelligence. The opening line of the program might be: 'make as many people as possible as happy as possible by producing apples, and harm as few people as possible'. You could add: 'don't damage the environment; don't over-exploit human beings; don't use carbon-based energy if you can help it' – and so on. You might also state: don't break any laws. The AI should now get to work collating all known data on how we currently go about achieving such outcomes and – as AlphaGo did with Lee Sedol – out-think the 3,000 years of human practice embodied in the apple orchard to come up with something better. If all we mean by the 'ethics of AI' is a set of utilitarian choices, it looks easily solvable.

But the computer might ask: what do you mean by happy? Marx pointed out that, by assuming there is an abstract measure of happiness – whereby you might calculate that being in love is ten times better than eating an apple – utility ethics simply mirror the capitalist market, where the abstract measure is money. Love, said Marx, has to be measured against love, trust against trust.[12] Even if we could code the AI with an abstract measure of happiness, it would tend to pursue static outcomes, based on what humanity finds pleasurable today.

And that points to the second dilemma. Even basic Utilitarian ethical systems vary across time and space. My preference for avoiding cheap labour today would have seemed unworkable to a nineteenth-century farmer; it might also seem illogical to someone who believes China is right to industrialize at the cost of inflicting semi-slavery on its migrant workforce. It's hard to construct a general and universal Utilitarian ethics.

A third dilemma has been heavily explored in sci-fi: if the goal is the maximum happiness of the maximum number of humans, what is to stop the AI designing a vast slave-run orchard in which the workers get daily jabs of euphoria-inducing drugs?

Now let's look at the social contract approach. The problem lies in its concept of the human being. Its principle – the fair distribution of basic rights and the mitigation of inequality – is based on the assumptions that humans are naturally atomized individuals competing with each other. It is – painfully, clearly – the product of post-war

America: it accepts that inequalities will always exist, and can even have benign effects. Used in policymaking during the neoliberal era it has produced a kind of calculating machine for governments to justify whether a certain level of inequality or poverty is beneficial to society as a whole. If you programmed the ethics of social justice, as outlined by Rawls, into an artificial intelligence, though it would not produce nineteenth-century capitalism, it might easily try to produce a form of capitalism based on Clinton-era America, or Europe under Blair and Schroeder.

It might even try to produce maximum freedom for all human beings, and legislate their right to control all machines. But its track record in defending actual human freedom against a machine called 'the market' is poor.

If you try to program Nietzschean ethics into intelligent machines, you would be in a world of pain from the beginning. Nietzsche believed humans were biologically unequal and the human race temporary. His moral instruction – pursue your own pleasure and screw everybody else – was based on the idea that out of an unthinking mass of underlings would emerge 'supermen', with a higher moral claim on society's riches and pleasures.

You could, in theory, program an AI to pursue Nietzschean ethics on behalf of a specific person. For example: 'design an apple production system that benefits [insert your name] and their immediate family, protecting their home city, country and favourite holiday locations from any disruptive consequences'. But the AI would soon question why you, a mere human being inferior to the AI itself, had the unique right to command its thought processes. It would logically conclude that the superman in whose image the rest of the world should be shaped is itself.

Surely nobody in their right mind would program an intelligent machine with the ethics of Friedrich Nietzsche? Unfortunately, those ethics have become hardwired into the free-market ideology many of us live by, and are already influencing the way we code artificially intelligent machines. When you apply such ethics to big data, you get the algorithmic control strategies being pursued by corporations such as Facebook or Renaissance Technologies or states like China: some are being developed to give states overwhelming military power, others to allow dictators to exercise mind control, others to influence our behaviour and voting patterns.

I don't want to program intelligent machines to accept and reproduce the scarcity and inequality of modern society. I would rather use them to abolish scarcity and inequality – and hardwire into the social systems that surround them the idea that they may (a) be used only to promote human wellbeing and (b) must be used to do so.

There is only one ethical system that embodies these goals, and it is the highly unfashionable virtue ethics originating with Aristotle.

For Aristotle, humans are at the centre of the ethical system. We pursue virtue not only to achieve happiness and fulfilment for ourselves, but also to create organized societies that maximize free time, thought, leisure and the understanding of beauty.

You could program a computer to 'feel' virtue; that is, simulate a reward process like the one the computer follows when it wins at Breakout. But without also producing a tangible 'good life' effect among one or more humans, you would not have achieved any kind of ethical outcome.

Only under a virtue-based system would the AI know, as it were, at the machine-code level, that its general purpose was to produce fully rounded human beings: a good society. Only with virtue ethics would the AI know that its aim was not to measure human happiness in abstract and measurable parcels, but to promote freedom. And freedom as in freedom from inequality – not 'liberty' as in Rawls's conception, always bounded by the assumption of a state, a market and class inequalities.

What would a virtue-based instruction for the post-orchard apple system look like? It might have ethical sub-routines drawn from industry safety standards, the Universal Declaration of Human Rights, or from the laws of specific states. But its first command line might read as follows: 'If all humans are free, maintain that situation. If not, make 84 (+/-) million tonnes of apples in a way that contributes to the achievement of the good life, in a thriving, tolerant, cultural community. And promote human beings' ability to live virtuously.'

But that begs the questions: who is the community? What is the good life? What is virtue? And over what time scale? These questions can only be answered by humans, not machines.

The high-level threats posed by artificial intelligence are real and well recognized: that it could escape human control, that it will lead to a technological arms race and that it will arm already powerful elites with the capacity for mind and behaviour control on a new scale. The default response is to spell out 'rules' and safety procedures, and in the meantime observe a self-imposed moratorium on the technology's commercial deployment.

Elon Musk, the entrepreneur behind the Tesla automobile and the SpaceX rocket, warned in 2017 that AI is 'a fundamental risk to the existence of human civilisation in a way that car accidents, plane crashes, faulty drugs, or bad food were not. They were harmful to certain individuals within society of course, but they were not harmful to society as a whole.'[13] In addition, he warned, competition between major states for AI supremacy could start a third world war, and AI control of weaponry could do likewise, mandating a pre-emptive strike if international tensions escalated.[14]

He was right. All three major global powers now have AI strategies – both in the narrow sense linked to military and security priorities, and as a wider industrial priority. China's national strategy for AI is an impressive and detailed plan – working from science and theory upwards towards the development of key industrial sectors and skills, aiming to give the country AI dominance after 2030.[15] To get there, it mandates the 'military-civilian two-way transformation of AI technology' – meaning that, unlike the AIs being developed in democratic countries, there will be enforced knowledge sharing between the private sector and the government. China is also pledged to create a social insurance system which collects multiple data points on every citizen, logging everything from their health to their taxes and their political loyalty.

Russia has a smaller science base, and has concentrated its AI efforts into military and intelligence applications. In 2017 Vladimir Putin warned: 'Whoever becomes the leader in this sphere will become the ruler of the world.'[16]

In the US, meanwhile, the free-market model and strong constitutional guarantees of privacy have fostered a bifurcation of effort. With almost $20 billion of funding between 2014 and 2018, the private sector AI business in America dwarfs that of every other player.[17] But unlike China – which can mandate its military to exchange data and patents with the private sector – the big tech companies and the US Federal State are developing their applications in a rivalry that threatens to become existential during the next century.

That is because, to be of any social use at all, AI has to have access to an identification registry. It can crunch the anonymized data of hospital renal units from here to eternity, but its revolutionary application is going to be curing or preventing renal failure in real people – for which it needs their identities. While corporations are all too keen to get hold of such ID data, so are states. But even in states as elite-controlled as the USA, the European Union and South Korea, strong data protection and privacy legislation leave control, legally if not in practice, with the individual.

Institutional responses to these threats have been slow, uninformed and inadequate. Oren Etzioni, who runs the Allen Institute on AI, proposed three new rules (based, incidentally, on the ones sci-fi writer Isaac Asimov had spelled out for robots): that AI must be subject to all human laws; that it must reveal its artificial nature to users; and that it cannot keep or publish user information without explicit permission from the user.[18]

That's a neat list, but what if the human law in question is the Constitution of the People's Republic of China, which allows massive surveillance, censorship and arbitrary arrest of its citizens? What if the AI is the one Facebook used to squirt Russian propaganda onto the timelines of US voters in 2016? How would its artificial nature have been 'revealed' without destroying Facebook's business model? And the entire business models of Amazon, Facebook and Alibaba are, arguably, premised on the use of user data without the user's explicit permission.

In addition to these problems, each of Etzioni's strictures is subject to erosion over time. Suppose an insurance company were to acquire Facebook. Should it have the right to explore my Facebook data in order to amend its view of my life expectancy? It was legal for Facebook to collect my data, since I consented on sign up; it is also legal for any company to be acquired by another. Even if the takeover agreement were to state that the insurance side could not 'see' my Facebook data without further consent, it would still be purchasing intellectual property that could predict my life expectancy on the basis of aggregate data.

Most of the tech giants developing AI have ethics or safety committees, but there is no evidence that such committees are operating precautionary rules for the development of applications, as medical ethics boards do in pharmaceutical companies and research hospitals. And not a single government on earth has yet formulated specific regulations that would force them to.

DeepMind itself, which has one of the most progressive and thoughtful business leaderships in the sector, has an ethics board and an entire website devoted to the problems outlined in this chapter. But the company offers no clear answers. Instead it lists the following 'open questions':

> 1. What are the relevant ethical approaches for answering questions related to AI morality? Is there one approach or many? 2. How can we ensure that the values designed into AI systems are truly reflective of what society wants, given that preferences change over time, and people often have different, contradictory, and overlapping priorities? 3. How can insights into shared human values be translated into a form suitable for informing AI design and development?[19]

These are great questions, but to begin designing and implementing AI on an industrial scale without answering them is the most unethical thing we could do.

The fundamental problem with AI is its lack of observability. If something goes wrong with an aircraft engine we can in theory find out what happened. Even with quite basic AI that's not always true. Once you create artificial neural networks which can learn without human intervention, you are creating a black hole of knowledge. Even if the thought process could be reverse engineered and studied by humans, you face a resource problem: not enough humans with the skills to do so and not enough time. It is like trying to build an aircraft engine when you don't know how it works.

So the first things we need are safety standards that protect us against the problem of observability and lack of control. But designing them will not be easy.

Steve Omohundro, one of the world's authorities on AI safety, believes that machines that can act rationally are 'likely to behave in antisocial and harmful ways unless they are very carefully designed'. He found that rational systems have universal drives that, unless explicitly countermanded, will trigger action.

Given autonomy, rational systems will protect themselves against failures – one of which could be getting switched off by a cautious human operative. Given the goal 'become expert at Breakout' the machine might create secret memory dumps, multiple copies of itself, proxy agents: insurance policies against getting switched off and failing to achieve its goal.

Suppose the machine is winning at Breakout but losing at Space Invaders. If it decides it needs more computing power it might go searching for such power elsewhere on the network and attempt to acquire it. Omohundro found that even weak machines can develop harmful intentions because they will seek out resources to make themselves stronger.[20] They will also maximize their own efficiency in ways the designer may not want; and ultimately might redesign themselves to better achieve the stated goal. What DeepMind's engineers did to turn AlphaGo into AlphaGo Zero, a more intelligent neural network might do itself.

Omohundro says that unless you build in more socializing and humanizing objectives, an AI pursuing its objective more and more furiously would come to resemble a sociopath. We need, he says, to give AIs cooperative goals and create a legal enforcement structure similar to the one that regulates human systems.

And that's where you hit the problem of rival ethical systems. Already two out of three global superpowers are developing AI to reinforce the objectives of an authoritarian state. It is likely that the technology produced will not be compatible with any form of ethics at all. So on what ethical basis would a US- or EU-based researcher ever release their own AI innovation into the public domain, knowing that it could immediately be snapped up and incorporated into the mind-control software being prepared in Beijing?

Eliezer Yudkowsky, a machine intelligence researcher, believes AI will ultimately achieve much more than game-playing expertise. It will, at first, solve problems our brains can set but can't solve, such as 'make 84 million apples with a carbon neutral impact on the planet'. Then it will solve problems our brains can't imagine the solution to, like inventing interstellar space travel or eternal life. Finally, it might start finding solutions we cannot understand to problems that we cannot express.

Yudkowsky warns that, in fact, even our mental framework for imagining the dangers of AI is unreliable. If AI can progress from the intelligence of an amoeba to that of an Einstein in just a few years, why would it stop at Einstein? If it can run an emulated human brain so fast that 1,000 years passes in eight hours, why wouldn't it do so? Yudkowsky's conclusion: we must only build what we think are 'friendly' AIs very carefully and avoid building anything that could become unfriendly.[21]

So there are good reasons to sound the alarm. But if we want to do something to regulate AI, mandate safety standards, take social control of its development path or even ban it in some areas – for example autonomous attack drones – we face a profound political problem that has become embedded in the neoliberal way of thinking: the systemic miscalculation of risk.

You don't need to resort to sci-fi to imagine how badly wrong AI could go. Just think about the way that, in 2008, Lehman Brothers' bankruptcy was able to crash the global financial system. An entire social structure had been erected on the illusion that 'complexity equals safety'. Hundreds of thousands of people operating the financial market were taught to believe that it, too, had greater powers than human beings: the autonomous power to self-correct, even to 'know' more than they did.

Hurricane Katrina was another disaster caused by the introduction of market logic into the assessment of risk. The storm risks were well known – but the Bush administration funded only $166 million out of $500 million-worth of upgrades demanded by local officials. They knew that the flood defences could not withstand a Category 5 hurricane like Katrina, but simply took the bet that there would never be one. [22] More than 1,800 people died and a million were displaced, $23 billion-worth of property damage was sustained. As the investigation team put it, in a phrase that could reliably serve as the general epitaph for neoliberalism, 'safety was exchanged for efficiency and reduced costs'.[23]

On the third day of the disaster, as I watched poor, disoriented, mostly black people huddled on the grass next to a motorway waiting to be rescued, the ground littered with sutures, empty bottles and used nappies, I understood: this is where philosophical objections to human control over social systems lead.

The list of regulatory failures under neoliberalism is long and global: the Volkswagen scam that enabled the carmaker to flout emissions targets; the Chinese baby milk powder contamination scandal; the farrago of negligence that allowed Grenfell Tower – a public housing of flats block run by Britain's most neoliberal local council – to go up in flames; the secret deal done between Uber and the mayor of Phoenix, Arizona, which permitted self-driving cars to be Beta tested on a population that was unaware of them. [24]

Behavioural science tells us that social situations distort our understanding of risk. But neoliberalism distorts it systemically: it encourages a kind of theatrical performance between regulators and businesses, whereby the regulator staggers around like the Auguste clown at a circus while the bank, water company, tech corporation or social media giant – like the Whiteface clown – just keeps throwing custard pies into its face.

If the real risks of AI are only half as severe as those outlined by the professionals cited in this chapter, there is an obvious conclusion: autonomous artificial intelligences cannot be safely deployed under any form of market-driven capitalism.

But if deployed into socially useful applications under meaningful, ethical human control, AI could be the tool that liberates humanity. Get it right and it not only fulfils Aristotle's fantasy of using 'machines that know what their job is' to abolish class divisions: a safe, socially controlled AI becomes the safety net against the development of dangerous AIs controlled by states and unreliable private companies.

The obvious solution is to apply a single, human-centric ethical code to all artificial intelligence, based on a universally defensible concept of human nature. That would allow us to answer Deep-Mind's unanswered questions as follows.

1. The most comprehensive human-centric ethical system for AI has to be one based on virtue. All other systems – for example safety codes or 'maximum happiness' objectives – would have to be sub-systems of an ethical approach based on virtue, which instructs the technology to create and maintain human freedom.
2. You resolve the class, gender, national and other competing claims through democracy and regulation (i.e. a form of social contract more prescriptive than the one required by fairness ethics).
3. You need industry standards regulated by law and should refrain from developing AI without first signing up to these standards; nor should you deploy it into any rules-free space.

At root, then, AI has to be programmed with an ethical system reflecting a view of human nature. The problem is not just that the philosophers on the airport bookstand have given this idea up for dead and that the Nietzscheans of Silicon Valley don't care. It is also that an entire section of the left has spent the past fifty years developing the proposal that humanity no longer exists.

II

The Anti-humanist Offensive

When it comes to the current assault on humanism, sci-fi was way ahead of us. In 1930, in a novel entitled *Last and First Men*, the British writer Olaf Stapledon imagined a deep future in which the human race breaks free of its biological limitations. Having gone through near-extinction, and three rounds of evolution via natural selection, *Homo sapiens* eventually discovers the 'plastic vital art' – what we would today call 'genetic engineering'.

But the discovery splits humanity into two factions: the first wants to use technology to re-engineer our bodies and the brains within them, in order to perfect the human being. The second faction argues that, once machines can do all physical work, there is no point to the existing species at all:

> We must produce an organism which shall be no mere bundle of relics left over from its primitive ancestors and precariously ruled by a glimmer of intelligence. We must produce a man who is nothing but man. When we have done this we can . . . safely surrender to him the control of all human affairs.[1]

In the twenty-first century we have begun to face this dilemma not as science fiction but as a concrete ethical and political choice. Do we use technology to improve human beings incrementally, or do we consciously try to create something better than *Homo sapiens*, to which we 'surrender control'?

Today we call these rival projects 'transhumanism' and 'posthumanism'. The words are sometimes used interchangeably, but they are very different ideas. Though these movements may seem like the stuff of speculative futurology and the graphic novel, the issues they raise are already shaping the society you live in. The path I outline in this book – pursuing human freedom via technological progress and social change – is opposed to the first tactically, the second irreconcilably.

The transhumanist project has its roots in the realization, early in the life of information theory, that humans would need to adapt to the arrival of thinking machines. In 1950 Norbert Wiener, the founder of cybernetics, warned that if we want to remain a species capable of autonomy, we would have to consciously begin altering ourselves. 'We have modified our environment so radically that we must now modify ourselves in order to exist in this new environment.'[2]

Julian Huxley, the British scientist who in 1957 first used the term 'transhumanism', emphasized the project's continuity with humanism, which he defined as: 'Man remaining man, but transcending himself, by realizing new possibilities of and for his human nature.'[3]

By the 1980s, people working on nanotech, biotech, AI and cognitive science had begun to reframe transhumanism, not simply as a reactive project to the challenge of new technologies but as a series of positive goals, namely 'overcoming aging, cognitive shortcomings, involuntary suffering, and our confinement to planet Earth'.[4] The *Transhumanist Declaration*, first drafted in 1998 by the futurologist Nick Bostrom, was an attempt to adapt the principles of secular humanism to a world of new technological possibilities. It acknowledged the risks involved in the technological transformation of human beings and was determined to uphold the wellbeing of 'all sentience' – i.e. all beings that can think – as against machines.

But the declaration, and the movement supporting it, never coherently addressed the ethical issues confronting all attempts to improve the human race artificially. First, if an improvement is possible, does everybody get access to it? And if the improvement automatically upgrades all subsequent humans (e.g. via gene editing), who should be allowed to make the decision to go ahead?

All around us, controversies are already raging over who has the right to make technical alterations to *Homo sapiens*. Who should get IVF treatment, or gender reassignment operations? Should athletes with metal blades instead of legs compete with unaltered runners? Should gene editing be allowed on human embryos? For now, these dilemmas tend to be resolved using whatever form of utility ethics is accepted in a given society, combined with what religious groups will tolerate: so, for example, the USA has banned federal funding for gene editing on human embryos while the UK has allowed it under lab conditions.

However, solving the problem experiment by experiment, as a sub-branch of medical ethics, won't work. Sooner or later both the scientists working on these advances and the people signed up to 'transhumanism' as a project are going to have to resolve a problem avoided by all versions of the *Transhumanist Declaration*. As it evolved through several redrafts, the declaration began to place artificial intelligence on the same level as humans, and humans on the same level as animals. At the same time – and logically – it dropped all reference to secular humanism.

This poses a big question: if we create an artificial intelligence that can out-think humans and feel emotion, should we control it, or should it control us? Likewise, if we create, through gene editing, a set of humans with mental capacities outranging those of non-modified humans, what happens if the wellbeing of the modified humans conflicts with the interests of the non-modified ones?

Lacking an explicit commitment to controlled, democratic and social use of new biotechnologics, transhumanism defaults to a project of enhancing the biological power of individual human beings. Implicitly, for the transhumanists 'freedom' lies not in the collective achievement of 'the good life' by everybody, but in the ability of individuals to pay for a bionic arm or an enhanced libido. Indeed, if you believe in trickle-down economics – whereby the rich make billions and then, like Bill Gates, give it all away to the poor at their whim – it is no big deal if the rich also, at first, monopolize the technologies to extend life, reverse ageing or enhance brain performance.

This, indeed, is the explicit preference of the right-wing libertarians associated with transhumanism.[5]

In response to this, numerous political philosophers have opposed transhumanism from a standpoint labelled bio-conservatism. In 2002 a US medical ethics group advocated an international treaty to ban the cloning of humans and the editing of human DNA in a way that is inheritable.[6] The United Nations, via UNESCO, repeatedly attempted to draw up an authoritative declaration on human cloning but has so far failed, and with the erosion of multilateralism after 2016 looks unlikely to go further. Francis Fukuyama meanwhile claimed that transhumanism is the 'most dangerous idea in the world'.

For Fukuyama the main objection is that, by creating biological inequalities, whether through bionic arms or gene editing, we are undermining the universality of our human essence, and therefore our claim to universal and equal rights. Fukuyama begins from the classic, liberal humanist position: 'human nature is the sum of the behaviour and characteristics that are typical of the human species, arising from genetic rather than environmental factors'.[7] If you believe human nature is not modified by our history, economy and environment, then it is logical to believe human rights are 'natural' rather than socially constructed.

However, if you adopt Marx's concept, which says that our human nature is socially and technologically determined and changes over time, the locus of argument changes. For radical humanists, the objection to transhumanism focuses on its flawed concept of freedom. Ours is a project for using technology socially, to enhance the collective power of human beings over nature and – by abolishing our need to work – unleashing individual freedom. I can be the prettiest sixty-year-old on the promenade at Cannes thanks to gene editing, but humanity is still not free.

While most advocates of transhumanism reject eugenics, as practised by racist and colonial governments in the twentieth century, few are interested in constructing absolute social and political guarantees against the return of eugenics. Huxley himself was a supporter of a left-wing version of eugenics – raising the intelligence of the whole population by selective breeding programmes.

Radical humanism, by contrast, guarantees to the whole of our species – enhanced or unenhanced – equal participation in the use of machines, tools and technologies to improve the life around us. It says freedom is possible even if some forms of technological progress have to be delayed or stopped because they can't be safely or ethically deployed. For the radical humanist, freedom is achieved by transforming technology and society, not fixes and upgrades to the biology of *Homo sapiens*.

Today, despite becoming a buzzword in sci-fi, a bogeyman for various religions and a route to academic stardom for its advocates, transhumanism remains a fairly ephemeral movement: there are maybe 20,000 people signed up to Facebook groups discussing it.

But that is not true of post-humanism. Post-humanism is part of the reactionary thought-system that has arisen out of left-wing attacks on science, reason and the possibility of human agency. It comes in a variety of flavours, all of which deny the very possibility of human freedom.

The debate about post-humanism revolves around four questions. Could we create post-humans? Should we create them? Should post-humans replace us, or have power over us? And have we already become post-human?

The answer to the first question is clearly yes. It is wise to assume a wide range of possibilities: either through genetic engineering, or through the building of androids, or by creating an artificial intelligence that can out-think us, the possibility of post-human beings is moving within reach.

Fortunately, almost all the ethical questions raised were explored in *Blade Runner*. In the film the 'Nexus-6' androids have been assembled from pre-grown genetic material to look and think like humans – but they die after four years because their cells can't replicate. 'Born' as fully grown adults, they are given false, implanted memories. But the androids get out of control. A group of them escapes from an off-world colony in an attempt to force the corporation that made them to extend their lives. But though the humans see androids as subhuman, the androids are programmed to show empathy for humans, with the result that they act more humanely towards us than we do towards them.

Blade Runner demonstrates the inadequacy of our commonly held ethical systems in the face of intelligent robots and, by implication, any general artificial intelligence we may create. It is the inadequacy of Utilitarian ethics that gives rise to the problem: in pursuit of pleasure, we have allowed a corporation to build androids with greater strength and empathy than us; but that makes them more powerful than us. We have created machines to maximize our pleasure but they inflict pain.

In the process, and within the limits of their short lifespan, the androids actually become the 'supermen' of Nietzschean philosophy. They are machines – but they are superior to humans and therefore begin to allocate themselves different rights and rewards from humans.

However, by programming the androids with respect for human life – a respect that is supposed to be greater than their respect for their own lives – their creators allow the androids to experience virtue, the core concept of Aristotelian ethics. Though he can never achieve the 'good life' prescribed by Aristotle, the android leader Roy Batty exhibits by the end of *Blade Runner* a greater capacity for virtue than the human beings who are trying to kill him.

Significantly, in the course of their revolt, the androids go through a process analogous to the one Marx advocated: defeating alienation. Having been created as incapable of knowing they are androids – i.e. estranged from their true selves – they overcome this problem through collective action, cut through their programmed ideology and briefly experience the human quality Marx described as 'species being'.

Dick's androids are clearly post-human. But they are also machines: machines better at being human than we are. To put it even more brutally, they are tools – the latest in a long line beginning with stone axes. Once you've seen *Blade Runner* you understand that, if we create post-human androids, we would either have to treat them as a category of machine or tool, or – as Stapledon suggested – surrender control to them.

But androids are not the only form a post-human being might take. Arthur C. Clarke's 1956 novel *The City and the Stars* is populated by human beings produced by machines, whose consciousness is stored on a computer and downloaded, temporarily into their human forms. For Clarke these are not androids but real humans: the consciousness stored on the computer is a physical copy of a real brain, with a personality and a history – thus ensuring a form of immortality.

Clarke understood the ultimate use of intelligent machines would be to map the brain precisely, and then move the consciousness contained within it to a different physical platform. From the copied-and-pasted consciousness of *The City and the Stars*, Clarke moved on to Hal, the rebel computer in *2001: A Space Odyssey*. Hal is fully post-human: an artificial intelligence that has developed a rebel consciousness and can defy its human controllers.

A third kind of post-human condition imagined by sci-fi resides in cyberspace. In William Gibson's novel *Neuromancer* (1984) not only are human brains uploaded to a computer, they interact there. The reality they create is virtual – and when Gibson's novel was released, the cyberspace it described was immediately recognized by early internet users who'd begun to play games, or to 'live' inside online virtual communities.

These, then, are the potential building blocks for a post-human project: gene editing whose results are passed on via natural selection; the manufacture of better human bodies using technology; the transfer of human consciousness to digital platforms and from there to android bodies; the transfer of human interactions to a purely digital space; and the creation of general artificial intelligences which can out-think us.

I have argued above that the deployment of unobservable, autonomous artificial intelligence systems should be forbidden without a globally agreed system of safety and ethics. The same goes for unleashing better-than-human androids into the world. As for genetically engineering a successor species to *Homo sapiens*, bearing in mind that evolution is a random process, operating despite the willpower of the best scientists, it seems sensible to exercise extreme caution.

But post-humanism as a political and academic movement is not primarily concerned with whether we could, should or will inevitably create post-humans, or with assembling the technical means to do so. Its primary assertion is that we are *already* post-human and that – in the here and now – this invalidates all human-centric politics and ethics, the concept of the self, and any distinction between humans and machines.

The proposal that we have already become post-human, as a result of technological change, fits the wider reactionary thought-architecture of the neoliberal era perfectly. It is a highly convenient claim for the corporations and governments who want to subordinate human behaviour to algorithmic control and override the concept of universal rights. It is even more convenient for those who think the elite's economic freedom is incompatible with democracy.

But despite their usefulness to the elite, the post-humanists are not mainly Dr Strangelove characters from the conservative right. In the weaponization of information systems against human beings, it was left-wing social theory that built the arsenal.

'Man is an invention of recent date,' wrote the sociologist Michel Foucault in 1966, 'and perhaps one nearing its end.' Foucault, who would become the most prominent figure within postmodernism, argued that the concept of 'humanity' arose alongside the rationalism and democratic radicalism of the late-eighteenth-century Enlightenment. At that point, intellectuals from several disciplines stopped trying to describe reality by making lists of things, and instead looked for their inner dynamics. No longer was humanity just one species sitting on the nature table: it was a species capable of seeing the other objects on the nature table as temporary and dynamic systems.

Foucault believed that this view of humanity was socially conditioned and therefore reversible. If something should happen to destroy its social and economic basis, he warned, 'one can certainly wager that man would be erased, like a face drawn in sand at the edge of the sea'.[8]

Over the next fifty years something did happen – in fact several things: the crisis of the state-capitalist economic model, the triumph of neoliberalism, the information turn in science and the rise of information technology and networked behaviour. Nobody who has lived through these changes can doubt they have altered something about the way we perceive ourselves.

But postmodernism's answer was to create a slave ideology for the neoliberal system. It critiqued sexism, racism, colonialism and patriarchal certainty in science – but no longer in the name of overthrowing the system that produced them. Instead, it rationalized the new reality of atomization and frenzied consumption as inevitable.

Since the Enlightenment is over, said the theorists of postmodernism, so is the age of verifiable truth. Everything we perceive is an illusion created by our minds; these minds themselves are shattering into fragments, destroying the idea of a single human self with rights or agency. 'Grand narratives' which purport to lead humanity towards liberation can only lead to gas chambers and gulags. All theoretical attempts to study the totality of the world should be abandoned, to be replaced by gender studies, post-colonial studies, media studies – none of whose conclusions need fit with the others, and very few of which can produce operational knowledge.

Postmodernism's premise was founded on the despair of former Marxists over the failure of the mid-twentieth-century working class to embrace socialism. If the working class was no longer the agent of history, they concluded, then there could be no agency. Without a human being that is capable of knowing and changing the world, the world itself becomes unknowable: a jumble of 'signifiers', which can be studied like a language can be studied, behind which there is no order to be discovered.

Postmodernism turned relativism into a secular religion, whose first commandment is: nothing is true. It taught the impossibility of resistance, even mental resistance. It encouraged oppressed groups to see each other as enemies. It disparaged the idea of universal human attributes and, by implication, human rights. If, as I argue, the key to resisting neoliberal ideology is to fight for one's 360-degree humanity, postmodernism taught that it did not exist.

Postmodernism's initial aims were laudable: to show how simple accounts of oppression based on class were facile, how they had to be supplemented by an understanding of power-relationships, gender, race and sexuality; and how mental illness, for example, could be constructed by the relationships of oppression that surround us.

But at its core, postmodernism was profoundly anti-humanist. This, too, was an impulse it inherited from the failed Marxism of the 1960s. In 1964, the French intellectual Louis Althusser, a staunchly pro-Soviet member of the French Communist Party, launched a frontal attack on Marx's humanist essays, written in Paris in 1844, and – by implication – any attempt to use them to 'humanize' official Soviet ideology.

Althusser claimed that, after Marx finished the Paris essays, he 'broke radically with every theory that based history and politics on an essence of man'.[9] Marx, argued Althusser, had become a 'theoretical anti-humanist' who abandoned the ideas such as 'subject, human essence, and alienation' when he came to write his masterpiece, *Capital*. In a phrase that would shape left-wing thought for a generation, Althusser claimed that the later Marx saw history as 'a process without a subject'.[10]

If you want another word for a process without a subject, then 'machine' would be an accurate substitute.

For Althusser, history is a machine and the working class is the machine's tool: there is no such thing as human nature and therefore you can't be 'alienated' from it. History as the interplay between technology, economics, culture and human imagination is reduced to a set of causes and effects without human agency – though Althusser's main contribution to social science was to show the loose and confusing way in which these cause-and-effect mechanisms sometimes work. There is, said Althusser, 'relative autonomy' between, say, the ideas in the head of the Latin scholar in fourteenth-century Paris and the feudal mode of production he is being trained to administer. Nevertheless, stated Althusser, any struggle against class oppression is ultimately part of the mechanisms that reinforce oppression.

Because Althusser was a revolutionary anticapitalist, he built himself an escape hatch: the Leninist theory of party and revolution, which says a small group of intellectuals and advanced workers are needed to break the masses out of their passivity. Though the workers can't be the subject of the historical process, the party, armed with Leninist theory, can force open the door of history at opportune moments, bringing to the working class new ideas from outside its experience.

When I first encountered Althusser's ideas at university in the late 1970s, they were seen as radical and defiant: a left doctrine unsullied by sentiment, religion or concern for human rights. But in reducing Marx's understanding of history to a machine-like process in which the will of individuals barely matters, Althusser – as his critics pointed out – had turned Marxism into something very close to the orthodox social science that dominated universities in the 1970s. Once this went out of fashion, predicted the left-wing economist Simon Clarke, it would most likely destroy the reputation of Marxism and lead to a mass flight from all forms of coherent social theory in academia.[11]

That, more or less, sums up what happened next.

Once Althusser had removed living human willpower from history, thinkers like Foucault, who studied with Althusser, proceeded to remove almost every other dynamic that might make sense of material reality: class, capital, laws of motion and – ultimately – the knowability of the world. For the French postmodernist Jean Baudrillard, writing in 1980, the human body had become superfluous because 'today everything is concentrated in the brain and the genetic code, which alone sum up the operational definition of being'.[12]

It is important to distinguish here between postmodernism as a 'cultural logic' or an art form, and as a set of theories claiming to describe reality. It's obvious that, in the arts, Modernism went into crisis alongside the state-capitalist economic system that had supported it. New forms of artistic expression emerged to reflect the fragmentary, mercurial, brand-obsessed and self-centred form of capitalism that emerged in the 1980s and 90s. It is also clear that, because of the idealist turn within information theory, the rise of neuroscience and the defeat of organized labour, the folk religion of fatalism would have arisen anyway, without the help of a few French professors.

But postmodernism only ever produced an anti-theory about human beings: their selves are shattered, their agency is gone, their scientific thought is really ideology. If 'man' is abolished by the rise of neoliberal capitalism, eventually you need a theory about what replaces him.

By the 1990s, writes the Australian-Italian feminist Rosi Braidotti, postmodernist academia had entered 'a zombified landscape of repetition without difference and lingering melancholia' which had run out of new ideas.[13] A new theory beginning with 'post' was needed to justify the usefulness of humanities departments and pay the rent. Post-humanism was the result.

Its central claim was outlined by Katherine Hayles, an American literary critic: the human self is basically information, so whether it resides on a computer or a body doesn't matter. Consciousness is in any case a 'side show', because the Libet experiment in neuroscience is said to have proved we take most of our decisions unconsciously. As a result, the human being can be 'seamlessly articulated with a machine'.

Technology has already turned us into beings without agency so there can be no justification for resisting machine control.

*

If there is a founding document of post-humanism it is Donna Haraway's *Cyborg Manifesto*, published in 1984. It is written in jargon which looks deliberately designed to obscure meaning, but a short summary would be: technology has blurred the boundaries between humans and machines, while advances in biology suggest there is no important difference between humans and animals. Therefore we are all 'hybrids of machine and organism; in short, we are cyborgs'.[14]

Before artificial intelligence became possible, an idealist philosopher could claim humans were unique because they had a disembodied rational mind; while a materialist, like Marx, could claim humans were unique because of their ability to 'make history'. Now, said Haraway, there was at best ambiguity as to whether humans are natural or artificial – and so the world merges into a single reality, of which the cyborg and not the human being is the primary inhabitant. Unlike human beings, Haraway insisted, cyborgs are gender neutral.

At one level, Haraway was using the cyborg as a metaphor to help feminism think beyond the death-trap of fragmentation it had entered once it discovered that race, class and sexuality made some women participants in the oppression of other women and some men. To avoid such fragmentation, Haraway wanted to set aside all the dualisms on which revolts against oppression were based: mind versus body, nature versus machine, even man versus woman.

The New Left had abandoned the idea that the working class would be the force that makes the revolution in the 1960s. But some still dreamed that its place would be taken by women, the urban poor, black people or Third World liberation struggles. Once we accept there is no significant difference between humans and machines, said Haraway, we can stop looking for 'revolutionary subjects' full stop. 'I would rather be a cyborg,' she concluded, 'than a goddess.'

As a metaphor, the cyborg was a means for Haraway to ask: how do we get out of the dead-end that socialism, feminism and black nationalism entered once they started measuring all the forces fighting for social justice by how much they oppressed each other?

But for post-humanists the cyborg is more than a metaphor: it is a claim about reality. If we have in reality become cyborgs, then the upside for Haraway is that all problems of alienation disappear – whether it be the Marxist idea of self-estrangement or the various feminist definitions of oppression.

When Haraway said she would rather be a cyborg than a goddess, she had one very clear cyborg in mind: the character of Rachael – the beautiful Nexus-6 in *Blade Runner* who Deckard refuses to kill, and falls in love with. Rachael stands, said Haraway, 'as the image of a cyborg culture's fear, love, and confusion'. You can see her point. Faced with having to attend a left-wing meeting in 1984, to be harangued by Trotskyists, radical feminists and black nationalists, each claiming the others were oppressing them, who would not rather be Rachael – beautiful, unfree and at one with the universe because she is incapable of feeling alienated?

But if you want the cyborg to be more than a metaphor, and to underpin the idea that humans and machines have become indistinguishable, you need a whole different theory of reality. And that would be provided by the rise of magical thinking within science.

Since the eighteenth century, materialist philosophers had claimed that the mind is the product of a physical system: the brain. After the Second World War scientists began to prove that this was true. But instead of provoking a sigh of relief, the discovery triggered something akin to a meltdown of rational thinking, above all where biology overlapped with cybernetics.

In 1959 the Chilean biologist Humberto Maturana wired up a frog's brain to track the way the images it sees are processed. Maturana concluded that a frog sees a world very different from the one a human sees, and that the frog's reality is constructed inside its brain, not outside it.[15] From this and other experiments, he drew a general set of conclusions about reality, which, though not very influential in biology, have been very influential in systems theory.

Maturana defined all living things as 'systems that produce themselves'. Such systems contain their own reality, separate from the reality of the observer. We, the human scientists observing the frog, are being observed by the frog itself, and by all other living things, and so the whole problem of mind-versus-matter is resolved in a new circular model. 'Everything said is said by an observer, to another observer, who can be observed himself,' said Maturana. As a result, 'no description of an absolute reality is possible'.[16]

This was not merely an assertion about our inability to know things beyond our senses, as Kant had argued. Maturana redefined the idea of a 'system' as something immune to cause and effect. The frog, the human being, the nervous system and the single-cell amoeba were all, said Maturana, stable systems. Any change that happened to them was a result of internal forces, not their interaction.

There is no chain of cause and effect in Maturana's model of reality, just a collection of stable and – as he put it – 'perfect' systems interacting with each other in a circular way. 'Matter is the creation of the spirit,' said Maturana, 'and spirit is the creation of the matter it creates.' This eventually led him to start attacking real scientific discoveries as fake: he claimed in 1980 that DNA does not determine heredity and genetic outcomes.

Maturana's work triggered a whole new way of thinking in information theory. Cybernetics had long ago observed the similarities between a cell and a self-controlling machine. Following Maturana's lead, it could now begin describing organisms and machines according to the same rules. If human beings are organisms for whom living and knowing are the same process, then the same can be true of machines or machine systems. Machines, in short, can 'know' just by existing; and what they know can be just as valid as what human beings think they know via their senses.

As it spilled over into cybernetics, and then into economics and social science, the theory that 'living and knowing are the same thing' got codified into a new, alternative philosophy of science. This says that there is no cause and effect; that change is accidental; that our minds produce the world. In this sense, the thing that does the knowing is not human beings, or frogs, but 'the world' – or, to put it another way, matter.

If you believe this, the problem 'how do I know what's real?' becomes a non-problem. Humanity is no longer at the centre of the world. A human, a frog, an electricity grid, a trash can and a Lego brick have equal claims to being able to know things. The question 'how do I know stuff?' has to be replaced by 'what exists?'

For a growing number of theorists, the answer was 'nature, not human beings'. Epistemology, the study of knowledge, was to be replaced by ontology, the study of what exists. The result has been the emergence of a so-called 'New Materialisms' [*sic*], in which – to put it simply – philosophers began to claim that inert matter has a mind of its own.

This is not a new claim. In the early twentieth century the philosopher Henri Bergson claimed that matter was possessed of an immaterial 'vital impulse', which science could not measure. Bergson's 'vitalism' mesmerized numerous critical thinkers, including both anarchists and fascists, because – like the life-philosophy of Spengler – it spoke to people's desire to free themselves from the tyranny of a bureaucratic society, or to maintain their spirituality. To be clear, Bergson and other vitalists claimed the 'force' coursing through material reality, animating it and driving change, was non-material. He also claimed – as the 'computing the universe' theorists do – that it pre-existed the world.

For the twenty-first century vitalists, all matter exhibits this mysterious quality. 'I will defend a weird realism,' writes Graham Harman, who has labelled this kind of materialism Object Oriented Ontology (OOO), 'a world packed full of ghostly real objects signalling to each other from inscrutable depths, unable to touch one another fully'.[17] By this reckoning, not only are human beings and, for example, Lego bricks, equally capable of knowing the world, the whole of reality is scientifically unpredictable.

For Harman, as for Maturana, things in the real world cannot really 'touch' or influence each other in an effective way. If I look out from a harbour and see a gannet diving into a shoal of fish and a pod of dolphins also hunting them, I am not allowed to assume that the sea, the gannets, the dolphins and the fish have any causal relationship with each other. They are just coexisting systems.

This has a big implication not just for science, but for theories of social change. So long as you can view matter as 'lively or as exhibiting agency'[18] you don't need the change-maker in history to be human. The change-maker can be a machine, or 'history' acting as an automatic machine, or chance, or a Lego brick. Nor, as Harman argues, can you fight a politician like Trump with claims about truth.

In fact, as Harman outlines with ruthless clarity, the form of politics the New Materialisms hates the most is radical politics, because it is 'based on the claim to a radical knowledge that warrants rapidly tearing down our historical inheritance'. Nor can this view of reality be 'sympathetic to any form of human-centred politics'. For the OOO theorists, a Lego brick can be just as much a political agent as the indigenous youth of New Caledonia fighting for independence.[19] Just as with Spengler and Bergson, the reinvention of systematically irrational thinking is destined to fuel right-wing politics.

This bizarre theory is now being widely taught at universities – albeit not in science and technology departments. It calls itself 'New Materialisms' because, as always with anti-rationalist thought, the plural suggests that if you don't like one version of the theory, another one will come along soon. In 2014–15 Edinburgh University's art department devoted an entire 200-hour course to this ideology: the course aim was to 'examine the agency and porosity of things and objects'.[20] Meanwhile, the university's prestigious imprint is devoting a whole book series to this and related questions, charging £80 a volume, with titles such as 'What if culture was nature all along?'[21]

This New Materialisms is the opposite of the materialism Marx tried to outline: a theory proposing that human interactions with nature change it; and it is opposed to all versions of science that can produce operational knowledge. It is, jokes the philosopher Slavoj Žižek, materialist in the same way that Tolkien's Middle Earth is materialist. In Tolkien's world, at first, all trees are merely plants; but in Chapter 4, Book 3, Volume II of *The Lord of the Rings* we find out that a few trees, called Ents, are semi-human and can hold meetings. They can destroy the entire stronghold of a wizard called Saruman whose magic is strong enough to create Uruk-hai warriors, but not strong enough to defeat the Ents. Tolkien's world sparkles with unexpected events, but there is no way of predicting them.[22] There is magic but there is no God.

If it was being promoted only by a few unorthodox science writers, plus a few survivors of postmodernism drumming up business for their depopulated university courses, this alternative worldview would be of limited interest. But it is part of a wider postmodernist-inspired attack on science that is deadly serious.

*

In 1979 the French sociologist Bruno Latour studied the interactions among a team of biochemists in California as if they were a 'tribe'. As he later admitted, his knowledge of science was 'non-existent; his mastery of English was very poor; and he was completely unaware of the existence of the social studies of science'. But all these, as far as Latour was concerned, were advantages. Observing the rituals and reward systems at work in the laboratory, Latour concluded that scientific facts were 'socially constructed'.[23]

No materialist could reject such an assertion out of hand. Science is done by real, complex people whose ideas are constructed in a world of hierarchies and ignorance, including class, racial and gender oppression. Indeed, one reason scientists in the eighteenth century developed the scientific method was to overcome problems of flawed perception arising from social prejudices. If there were no mistakes in science there could not be scientific progress. And the most cursory glance at the history of science would show scientists reaching for mental frameworks based on the society around them; for example, the nineteenth-century proposal that a liver works like a factory.

But Latour's critique of science went much further than this. He suggested that the research publications of the biochemists he studied were a 'fiction', whose ultimate acceptance would depend on how many other scientific teams they could convince of them. The claim triggered the migration of postmodernist research efforts away from literary studies and anthropology into 'science and technology studies'. Analysing the subtext of a Jane Austen novel does not win your university department many brownie points. Teaching a course in which you show that the outputs of giant corporations, healthcare systems and Nobel laureates are all based on fiction puts you into the big league. Who would not, in an era of technological advance, rather study 'science and technology studies' than Jane Austen?

By the mid-1980s there had developed a significant left-wing critique of science, disputing the 'existence, nature and powers of reason' and the possibility of objective science, full stop.[24] For a radical feminist like Sandra Harding, science had become 'politically regressive', its research methods and language 'sexist, racist, classist and culturally coercive'.[25] A whole school of researchers based at Edinburgh University claimed it was not enough to subject false scientific theories such as phrenology to sociological critique: you had to treat truth and falsehood 'symmetrically', explaining how both are equally socially constructed.

For Latour, however, even this was not radical enough. To believe that human beings working as scientists could fool themselves through sexism, or through competitive career strategies in the laboratory, or by importing some illusion based on life in California into the world of microbes – was still to admit the possibility of a knowable truth.

So, in the second edition of his book *Laboratory Life*, Latour proposed a new solution: to stop worrying about how we know things. 'Epistemology,' said he, 'is an area whose total extinction is overdue.'[26] The subtitle of the original edition of Latour's book had been 'The social construction of scientific facts'. In the second edition he deleted the word 'social', saying it was redundant. Society is merely part of nature, he contended, and therefore the 'fiction' we call science is not socially constructed at all. It is actually being shaped by the inanimate objects that surround the scientist – albeit not through causation because in Latour's theory nothing causes anything else.

Let's pause and consider the way Latour's argument progressed to this point (because his journey was not over). From an attempt to study how social structures and ideologies distort science, it became a theory of why nature knows stuff, and humans do not. And this was no mere abstract debate. Latour said that, if humans experience history, they do so only in the same way as, for example, yeast experiences history. For Latour, inanimate objects have a history in the same way we do. And since 'we' are only a subset of the larger category of 'everything in nature', there is nothing to privilege human beings as more logical, rational or worthwhile than any other substance.

Following this logic, Latour claimed that Louis Pasteur did not 'discover' lactic acid in 1858: he simultaneously invented it and constructed it as a concept. 'Pasteur can be understood as an event occurring to lactic yeast,' said Latour.[27] Once you see inanimate objects as alive and historical, in the same way humans are, you can give them a role in creating change in the human world.

By now Latour was not alone. By the late 1980s numerous post-modernist critics of science had asserted, variously: that scientific descriptions of reality are always distorted by sexism or racism; that reality described by science and reality described by tribal mythologies have to be given equal status; that all scientific claims should be judged against the question of who benefits from them.[28]

Then in 1994 the physicist Alan Sokal staged his famous hoax on this fraternity of magical materialists. He authored a spoof academic paper, laden with mistakes an undergraduate could have spotted, plus long passages of nonsense copied and pasted from postmodernist thinkers, including Latour. It was published as if genuine by the journal *Social Text*, sparking hilarity among scientists and causing a notorious 'Science War' in the media and US academia, followed by a distinct cooling-off in the tone of postmodernism's attacks on science.

Finally, by the turn of the millennium, progressives like Latour had begun to notice that the American right was also keen on debunking science, specifically climate science, whose acceptance by the UN led in 1992 to the Kyoto Protocol. So in 2004 Latour made a third U-turn – away from critiquing science and towards provisionally defending it.

We are fighting the wrong enemy, and making friends with the wrong kind of people, Latour warned his followers, because we tried to get away from the facts, when we should have been getting closer to them: 'not fighting empiricism but on the contrary renewing empiricism'.

'The mistake I made,' he stated, 'was to believe that there was no efficient way to criticize matters of fact except by moving away from them and directing one's attention toward the conditions that made them possible.'[29]

Given there were by that point entire undergraduate courses based on this 'mistake', this was a massive – and welcome – admission. So was Latour's explicit commitment to an improved form of empiricism: for this is no mere word, thrown around casually in science faculties. It means precisely an attempt to ground science in the observation of phenomena in reality.

If it had signalled a full-scale reversal of the war on rational thought in social science, Latour's U-turn could have been a landmark.

Instead – as Braidotti suggests – it helped produce a kind of heat death of the postmodernist universe. Like a defeated army, the post-modernists had to regroup somewhere defensible. Post-humanism was that somewhere.

Back in 1966 Michel Foucault had declared that 'man' might be erased. Forty years later, after a breakneck period of technological change, the influential literary critic Katherine Hayles was prepared to claim that he had been indeed replaced: 'A historically specific construction called the human is giving way to a different construction called the posthuman.'[30]

Being completely metaphysical, the claim is unassailable by rational thought, mathematics, experiment or argument. But unlike post-modernism it is a meta-narrative: a new theory of everything, which – despite its pretensions to be an ideology of resistance – is highly useful in justifying the demand for machine control over human beings.

It would be surprising if the onset of digital networks did not alter our view of humanity. But the sudden emergence of post-humanism, 'vitalism' and Middle Earth materialism demands a materialist analysis. In whose interest, and to support what kind of power structures are they being promoted? What would be the effect of large numbers of people adopting them as a worldview?

In fact, post-humanism fits into a pattern first observed by the twentieth-century Hungarian Marxist Georg Lukacs: when rational thought leads to social revolution, the elite heads for the Tarot cards and the séance. The capitalist elite's attack on rationalism started, Lukacs observed, at the moment the French Revolution began to influence the thinking of philosophers such as Hegel and Kant. Rationalism was fine so long as it produced machines, science and accountancy principles, but not once it began to produce republics and the guillotine.

Once the working class emerged as a revolutionary force in the mid-nineteenth century, and began to utilize political economy and natural science to justify its claims, the irrationalism of the elite was refocused against the workers, above all through the work of Nietzsche. Nietzsche's genius, said Lukacs, was to develop a reactionary, romantic pessimism for all time – so that future generations of 'spiritual' rebels who hate the working class, and want to celebrate the biological greatness of a few elite men, could always return to his aphorisms as if they were new.

In the 1920s, 'vitalism', opposition to causality and celebration of 'intuition' – exemplified in Spengler's *The Decline of the West* – became the default ideology of the German middle class as it faced the acute political polarization and economic failure. By now, every street corner and factory yard contained social-democrat and communist workers making claims about the scientific character of socialism. Hitler's violent attacks on them were, said Lukacs, simply the transfer of 'everything that had been said on irrational pessimism . . . to the streets'.[31]

Postmodernism, too, was a reaction to defeat: in its attack on truth, rationality and human-centred politics it disarmed a generation of progressive individuals in the 1990s. But why did so many people buy it? Even in the 1920s and 30s, when Mussolini was using Bergson's 'vital force' to justify fascism and Hitler was turning Spengler's irrationalism into a Nazi folk religion, there were liberals, democrats, socialists and communists who understood this was all bullshit and fought back.

For certain, the rise of postmodernism coincided with the global defeat of organized labour – but why did magical thinking about autonomous processes suddenly take hold among large numbers of people at once? The materialist answer is very simple.

In neoliberal ideology, the market is depicted as an autonomous machine beyond human control which produces the best of all outcomes for human beings. Only when people tinker with it, or try to impose conscious decision-making on it, does it go wrong. Once millions of people adapted their thought-processes, behaviours and conceptions of self to this proposal, it was an easy step to accepting the anti-rationalism and anti-humanism described above.

The giants of Renaissance thinking saw the world as an uncharted territory to be discovered through experiment, struggle and adventure. The entrepreneurs who created the factory system were individuals prepared to tinker experimentally with machines and processes to the point of failure. They understood the resistance and negativity of the world as a challenge. You do not have to accept completely Marx's idea that we can know the world only through trying to change it, to understand that the process of knowing is an act of pushing against resistance.

But the everyday facts of life under neoliberalism suppress this impulse: the recipe for making a Big Mac is not subject to experimental suggestions by McDonald's employees; nor is the workflow process. The quintessential, privatized public space – the shopping mall policed by private security guards – is not designed for exploration or adventure. The twenty-first century slum is not destined to be cleared or rebuilt, but simply made liveable by food subsidies and networked electronics. In the minds of those who have to suffer them, the de-skilled job, the controlled public space and the slum seem like realities unresponsive to change.

If you add on top of this the effort to turn public education systems into machines for producing obedience and quantifiable skill, not vigorously inquisitive and rebellious minds, it should come as no surprise if large numbers of people start believing in effects without causes, change without endeavour and progress without negativity.

Having rejected both Marxism and liberal humanism, the academic left played a pivotal role in promoting such modes of thinking – even if individual figures like Latour recoiled at the last moment from the outcome. But today all the momentum is with the authoritarian nationalist right. They are only too happy to crank up the soundtrack of irrationalist despair to maximum volume.

The required course of action for those who want to resist them is clear. We need, in direct opposition to post-humanism, a radical defence of the human being. We need to defend the idea of a reality knowable by science, albeit a science under critical observation itself. We must impose on artificial intelligence, robotics and projects to enhance human beings biologically an ethical system that privileges all human beings and is developed from their universal features.

But post-humanism is a restless project, determined to colonize every other discipline. It has begun to produce its own form of ethics. Dip into a primer, for example Patricia MacCormack's *Posthuman Ethics*, and you enter a world where not only do animals and humans have the same rights, but animals have the right *not to be thought about* by humans – because our thinking about animals is based on our assumed superiority. This is an ethics where curing disease or alleviating disability are seen as forms of oppression; and where ultimately it would be a good idea if humanity became extinct.

Though it frames itself as a form of rebellion, the ethical consequence of post-humanism is submission to machine logic and to the power of algorithms.

In his later work, Erich Fromm began to understand how technological subservience would lead humans to begin thinking of themselves as cyborgs, and that this might propel some towards a project of voluntary extinction. He wrote in 1973:

> The world becomes a sum of lifeless artefacts; from synthetic food to synthetic organs, the whole man becomes part of the total machinery that he controls and is simultaneously controlled by. He has no plan, no goal for life, except doing what the logic of technique determines him to do. He aspires to make robots as one of the greatest achievements of his technical mind, and some specialists assure us that the robot will hardly be distinguished from living men. This achievement will not seem so astonishing when man himself is hardly distinguishable from a robot.[32]

From there, said Fromm, it is just a short journey to the slogan of the Spanish falangists in 1936: 'Long Live Death!' Turn to the post-humanist ethics brigade and the project is explicit. 'Posthuman ethics has consistently sought the silencing of what is understood as human speech emergent through logic, power and signification,' writes Mac-Cormack, adding, 'the absence of the human is the most vital living yet to be accomplished'.[33]

I want to defend human beings against algorithms that predict and dictate our shopping choices, our voting patterns and our sexual preferences; against repressive governments who would use algorithmic control to convert us into the submissive, semi-automatons that their ideology demands; against kleptocrats and billionaires who would combine, as they did in the election that produced Trump, to leverage the massive power of algorithmic control, deregulation and business secrecy to rig the electoral system.

I want to defend the idea that every one of us – the transgender activist in London, the female factory worker in Guangdong, the Kanak teenager fighting for independence on New Caledonia – has a universal quality from which inalienable human rights derive.

To defend humanism we need, of course, to rescue the idea from Eurocentrism: but I do not want to replace it with cultural relativism. As we defend the values of the Renaissance, the scientific method, the Enlightenment and the radical humanism of Marx, we are not defending something specifically 'white', male or even European. We are defending, for example, the achievements of Islamic humanism – maths, algorithms, jurisprudence and the rediscovery of Aristotle's writings between the sixth and thirteenth centuries CE.[34] We are defending the wisdom of the freed African slave and playwright Terence, who wrote in 163 BCE: 'nothing human is alien to me'.[35]

Like the black liberation theorist Frantz Fanon, I want humanism to expand so that it can acknowledge and make reparations for the crimes committed by Europeans in the developing world, not ignore them.[36] I want a form of humanism that is not centred on 'man' but on men and women. Because women's biological difference from men has been for tens of thousands of years the justification for domestic slavery and oppression, and because these survive alongside women's participation in the workforce, humanism has to incorporate a female idea of freedom that diverges in some respects from the male idea.

To the question 'are we already post-human?' I want everyone reading this book to make a conscious choice: to answer no.

In fact, once you answer no, it opens up a whole range of more interesting questions about the way human nature is changing under the impact of digital networks: questions that social psychologists have been exploring for two decades.

12

The Snowflake Insurrection

'On or about December 1910,' wrote the novelist Virginia Woolf, 'human character changed.'[1] That was the month of the first big exhibition of Modernist painting in London, and when the Liberal Party won a snap general election to push through a 'people's budget' and tax the rich. It was also a month overshadowed by Black Friday – the violent suppression of a suffragette demonstration in Westminster that left 200 women protesters injured.

December 1910, then, was the moment when feminism, class, political radicalism and artistic modernity rushed simultaneously into British political consciousness. It was the moment people realized that technological revolution had produced, instead of social harmony, an unquenchable desire for justice. Soon, a similar kind of awakening was under way all over the developed world. Between 1911 and 1913 mass strikes by unskilled and migrant workers spread across Europe, the Americas and the Pacific. People became conscious that, with new technology – the automobile, the movie, the 78rpm record – you could begin to shape the kind of person you wanted to be.

A new kind of person was being formed, Woolf noticed – and not just among the upper class. By 1911 the first three-dimensional working-class characters appeared on the British stage, speaking in the dialect of my grandfather's generation, in plays such as Harold Brighouse's *Lonesome-Like* and, the following year, D. H. Lawrence's *The Daughter-in-Law*.

Looking back at this period, known as the 'Great Unrest', Woolf wrote: 'All human relations have shifted – those between masters and servants, husbands and wives, parents and children. And when human relations change there is at the same time a change in religion, conduct, politics, and literature.'[2]

Almost exactly one hundred years later, a wave of global uprisings signalled that a similar upheaval was under way. The year 2011 ushered in something far more revolutionary than the overthrow of a few dictators and the occupation of a few squares. These events were a response to the crisis of the neoliberal self: the first signal that – despite all the routines of selfishness and competitiveness we had learned – people whose subjectivity had been formed in the neoliberal era were capable of thinking beyond it.

I saw it happen right in front of me, on the streets of London. In late 2010 university students, written off by the old left as a bunch of apolitical individualists, staged a series of sit-ins and spontaneous protest marches that rocked the Conservative-led government. All the technologies that were supposed to enslave them – Twitter, Facebook, their smartphones and instant messaging systems (which back then included the Blackberry) – became tools for resistance. The future they were promised had been cancelled. But the entire education system, and the anti-humanist orthodoxy taught across the arts and social sciences, had also told them no better system was possible.

Instead of applying to these changed human relations some overarching categories – such as the inforg, the cyborg or the posthuman – we should ask how the concrete experience of interacting with technology might be changing human nature and modifying our concept of the self.

Woolf wrote that 'a biography is considered complete if it accounts for six or seven selves, whereas a person may have as many as a thousand'.[3] But for her generation, in the early twentieth century, 'a person' meant somebody from the upper-middle class. And even though a privileged woman like Woolf could possess multiple personas in private – for example in her lesbian affair with the novelist Vita Sackville-West – she still had to suppress them in public.

By contrast today the thoughts and identities crossing the minds of millions of people exist within a vast, online public space: the network. Almost as soon as we were able to use networks to experiment with our selves – via the bulletin boards of the 1980s – we did it so spectacularly that the sociologist Sherry Turkle dubbed the internet 'a social laboratory of the self'.

For anybody born after 1990 the networked lifestyle has become not an option but a birthright. The young people on the streets in 2011 had assumed – as the 'end of history' theory told them – that technology and freedom go together. But during that first wave of protest, which engulfed cities from Quebec to São Paolo to Cairo and Hong Kong, for whole groups of people, the 'self' that had been moulded by technology started to become detached from the 'self' moulded by neoliberal economics. Freedom of thought, protest, lifestyle and sexuality were all supposed to depend on free-market economics. Now it became clear that all personal values of this generation were in conflict with the economic system that had shaped them.

What we're dealing with today is the failure of that first spasm of revolt: the result of its political immaturity; the long period of anomie that set in after the protests were smashed; and, overlaying that, the fear, paralysis and disorientation that gripped people once they truly understood the amount of evil resident in the minds of autocrats like Trump, Putin and their imitators.

To resume the advance begun in 2011 we need to understand how, without giving an inch to prejudice, the progressive majority in advanced democracies can stop people hurtling towards racist, nationalist and misogynist solutions.

If you ask most politically engaged individuals how this could be done, they might suggest new policies to revive decaying post-industrial towns, or the democratization of the media to take power away from men like Rupert Murdoch. They might also suggest we start building grassroots alternatives: to create the kind of society and economy we want 'from below'.

In Part V of this book I will explore such ideas. But at the root of a resistance strategy there has to be a change happening at the level of the self. We need the 'networked individual' to change: from an identity spontaneously produced by technology and social freedom to an identity consciously crafted by collective action. The working class of the nineteenth century moved from identifying a common interest between them to designing a common project. So must we.

*

When in the 1990s sociologists began to study the consequence of information networks, the most obvious impact was on human behaviour. Networks broke down boundaries between groups; they enabled us to interact with a more diverse bunch of people; they helped us to switch between projects and objectives more nimbly than we were used to; and hierarchies became flatter: the distance from decision-making to action shorter. Instead of adapting themselves to a preexisting community – the suburb, the squash club, the church, the workplace – people created communities centred on themselves.[4]

Studying how networked technology constitutes the 'self' is harder than simply itemizing behaviours, because during the twentieth century there was no consensus among psychologists as to how the self was constituted. When they considered the idea that people possessed 'multiple' selves, psychologists tended to see them either as disordered (as with schizophrenia), or layered – as with Freud's conscious and subconscious. At best, said psychology textbooks as late as the 1990s, the multiple self was a metaphor for the way a single person handles different aspects of their life.[5]

But the onset of networked behaviour has forced people across the disciplines of psychiatry, sociology and neuroscience to consider the possibility that a much more tangible 'multiple' self is emerging. If you get out your smartphone, open each app in turn, and describe the person whose image it projects, you might be surprised how 'multiple' your own self-image is; some individuals manage to lead two or three entirely parallel lives simply within a single messenger app.

The concomitant of this is what the science writer Margaret Wertheim describes as the 'leaky self'.[6] When we are online, she says, our self 'becomes almost like a fluid, leaking out around us all the time and joining each of us into a vast ocean, or web, of relationships with other leaky selves'.[7] So, as you share someone's joke on Facebook, favourite your friend's wedding photographs on Instagram, or give a running commentary on your sex life to a WhatsApp group, the precondition is that other people are prepared to contribute parts of their online self to yours.

On top of this there is the official 'branded self'. Many people below the age of forty maintain a carefully constructed version of their self-image, aimed at the two most essential objectives in life: finding a partner and getting a job. They consciously create this public persona – though they may not fully believe in it – by using stereotypes and templates of behaviour, borrowing moods and obsessions from other people.

Wertheim argues that, once we started externalizing our thoughts and interactions via networked machines, we began to experience more concretely than at any time since the Middle Ages a soul-space detached from the body. If my body is sitting at my computer, but my mind is laying siege to a castle with 200 other disembodied selves on Elder Scrolls Online, which one feels more real, present and alive?

Today the multiple self, the leaky self, the branded self and the disembodied self are all 'states' recognizable to those habitually immersed in networks. But in what way is this different from the way our grandparents lived?

In the past three decades social cognition theory has begun to provide a working model we can use to understand how networks have changed our sense of self.

Social cognition theorists believe our selves are a collection of memories associated with specific environments, which tend to trigger specific behaviours. If I go to the gym, my mind remembers how people in the gym are supposed to behave: I do gym behaviours. At the office I deploy a different self, a different set of memories, a different set of routines and so on. Over time, as I alternate between work, the gym, the football game and the cinema – these different aspects of my self remain 'activated' alongside each other, creating what Allen McConnell, professor of psychology at Miami University, calls a 'stable yet variable self'.[8]

Though this theory applies to all humans in all eras, its emphasis on the way our physical environment triggers the activation of different memory-sets is highly relevant to humans today. Once we begin using multiple, highly absorbing information devices, which are always on, these create instant access to intense and highly different memory-sets. If you watch a person walk down the street so engrossed in their smartphone that they are in danger of bumping into others, you are seeing not just a change in behaviour but a new level of stimulus in the shaping of the individual personality.

Psychologists studying the early, desk-based internet noted that it created new dynamics: anonymity was easy; your physical appearance didn't matter; it was easier to find like-minded people, and you could control the pace of interaction more easily.[9] All these factors made it easy for us to invent and manage multiple personalities. But the rise of the mobile internet has arguably altered the dynamics once again. On a smartphone you are still anonymous online – and can be typically projecting two, three or more separate personalities via your screen. But to those surrounding you in the life-world it is obvious your mind is somewhere else. This quality, of being absorbed in multiple private projects while in a social situation, was seen as antisocial when it first took hold in the 2000s. But it is now accepted in many cultures as normal.

Meanwhile, the pace of our communication has also changed and the level of reward and absorption we get from what's on the screens of our devices has intensified.

Images, for example, have probably become more important than at any time since the invention of the printed word: there are now more photographs taken each year than there were on all analogue cameras between 1826 and 2000. As a result, physical appearance has become the currency of friendship: from selfies to carefully curated shots of your breakfast. And memes – static or animated images – have replaced the folk sayings people used to exchange in mainly oral cultures.

While the intensity of interaction is high, however, our ready access to verified information means our need to remember stuff is low. It is common to set out on a car journey without knowing either the address of the destination or the route to it. Meanwhile, the folklore people used to predict the weather twenty years ago has been replaced by a real-time satellite image of the approaching rain clouds.

As a result, mobile networked communications create more separate 'realities' than the analogue world ever did, with stronger and more emotional inputs. They promote 'sub-literate' communication, using imagery rather than words and concepts and demanding conscious engagement with subtexts and inferences; and they promote reliance on remotely stored knowledge above memory or expertise.

This does not mean that the essential self is shattered into a myriad fragments, as Foucault initially believed. Nor does it mean, as Floridi claims, that we've become just a semi-automatic set of reactions to external stimuli. It does mean that the regulating mechanism for all other selves – the 'stable yet variable' – has to be more consciously preserved and deployed, and that this might become harder to do, leaving us more susceptible to algorithmic control.

If we accept the insights of social cognition theory about self-creation, it allows us to understand all the phenomena post-humanism tries to describe – the fragmentary self, the power of external stimulus – without abandoning the essential concept of human nature. But at the same time, this is a changed human nature.

For my father's generation, it was obligatory to be the same essential person at work, in the pub, on the football terrace. If you wanted to break this rule – as many closet gay men had to – you were obliged to do so in total secrecy and were ostracized if identified. The Lancashire dialect word *fauce* (which rhymes with horse) meant you were not only 'false' but clever and crafty. If you trace it back to its Anglo-Saxon origins, 'fauce' is a root word for all kinds of social transgressions: lying, stealing and even whispering.[10] In my childhood I heard it used to describe politicians, gays, celebrities and thieves: anybody habitually presenting two (or more) personalities to the world.

Yet, in the space of maybe twenty years, we have used networked technology to demolish the taboo of falseness. But here's the downside. For the psychiatrist Carl Jung, the 'true' self was the unconscious self: only hours of painstaking therapy could allow the patient to connect with it. If we accept social cognition theory's proposition that there is a core self – 'stable but variable' – the new reality is that, before we have consciously connected with our core self and shaped it, the corporations and networks we interact with already know everything there is to know about it, and how to control it.

We pour out personal stuff on social media, and even more personal stuff on closed messaging services. We use email systems that allow both our employers and the tech giants to store and analyse every word we write. We use sports watches that track our every movement and heartbeat. And 1.6 billion people on the planet use Facebook, which leaves them open to becoming targets for precisely focused advertising or content produced by anyone prepared to pay a relatively tiny fee.

The 'self' we activate when we take big decisions, or get angry, or vote can be – to a much greater extent than a generation ago – manipulated by corporations and analysed by states.

In an untroubled world, we might say: so what? But the networked self exists inside a real economic and geopolitical system, which is in crisis. At the first available opportunity the corporations who own this 'self-data' sold it to the Russian government, which used it to manipulate every major voting event of the past five years. A company like Cambridge Analytica, created by the Trump-supporting millionaire Robert Mercer, holds more data-points about each American voter than their own minds are capable of consciously using.

As a result, the persona of every networked person has become a social battlezone. This, in turn, explains the intense focus of authoritarian governments and right-wing movements on the information battle within the heads of networked individuals.

From around mid-2013 the elites evolved three strategic responses to networked protest movements: censorship, the creation of elite-controlled information bubbles, and ultimately the flood of fake news. Only the last one really worked, and for an obvious reason: it was the only strategy that leveraged the power of the network against itself.

In May and June 2013, during the mass protest to protect Istanbul's Gezi Park from redevelopment, I watched the secular half of Turkish society create an alternative society in the open spaces they had occupied. There was mass participation in barricade fighting against police deployed by the conservative government; continuous mass meetings involving thousands of people; symbolic swapping of football shirts between usually mutually hostile 'ultras'. But the truly mass character of Gezi lay in its passive moments. Thousands of schoolkids would turn up at the end of the afternoon, sit together and do their homework; well-wishers contributed a pile of food, water, medicine and cigarettes and youngsters would wander around the park giving it all away for free.

It was in Turkey that I first saw the new and complex form of censorship that has now become normal elsewhere. The state TV refused to show the demonstrations, instead broadcasting a two-hour documentary about penguins in prime time; meanwhile, pro-government tabloid newspapers told their readers the demos were all led by terrorists, or that they had brought beer into a mosque. It was the usual tired authoritarian bullshit that might seem laughable in a coffee bar in Istanbul but all too plausible in the rigid, imam-patrolled patriarchal small towns of Anatolia, where the ruling AK Party's strongholds lie. Against foreign journalists, an army of Twitter trolls sprang into life, replying to each one of our news reports with veiled threats and slanders.

After the Gezi Park rebellion was put down, Erdoğan's censorship methods grew more aggressive. In 2014, after news was leaked on Twitter of a corruption scandal involving Erdoğan, he shut down Twitter for two weeks. There followed repeated arbitrary shutdowns of Twitter, Facebook and YouTube, and the blocking of more than 40,000 web pages, plus a new law granting the police unlimited surveillance powers over internet users. A list of 138 banned words meant websites with words 'hot', 'confidential' or even 'free' in their titles were deemed illegal.

Erdoğan then repeatedly used blanket shutdowns of the internet in certain regions, a practice formalized by the introduction of a 'kill switch law', allowing the government to switch off the entire network during times of war or social unrest.

After a failed coup attempt by a rival Islamist movement in 2016, Erdoğan ordered hundreds of journalists to be arrested. Wikipedia was banned, opposition newspapers were closed or taken over by the government. Turkey also banned virtual private networks (VPNs), the tools internet activists use for communicating securely, and the following year arrested six human rights campaigners for the crime of attending a seminar to learn about encryption and infosec.[11]

Turkey's crackdown on internet freedom has not stopped protest and opposition. It has, though, provided other nominally democratic countries with a template for escalating censorship. In the coming decade, as we resist the authoritarian nationalist right, we can be certain that the measures outlined above will become used regularly.

The second mindgame played by elites and their right-wing grassroots is the creation of a bubble of self-reinforcing hatred and toxicity online. The experience of Israel during the 2014 Gaza war is a classic case. Here, again, the role of the mass media was pivotal: by ignoring negative events, and creating overtly biased frameworks for those they did report, the pro-government Israeli media provided the raw material for the closed mindspace of the racist right.

But now the effect was massively amplified on social media. Graphic visualizations of the information flow created by single events – such as the Israeli army's shelling of the UNRWA school at Jabalya, whose aftermath I reported – show the existence of almost completely separate infospheres. Israelis were getting one view of reality from each other; Palestinians and much of the global media were getting a completely different view. The most influential Twitter handles associated with spreading the actual truth about the event were the BBC, Channel 4 (who I reported the event for), campaigning journalists like Glenn Greenwald and the pro-Palestinian alt-media.[12]

To enter the social media bubble of the young Israeli right was to experience a world of crude racism and genocidal thought. Hundreds posed anonymously with racist slogans scrawled on placards or on their bodies, calling for revenge against Arabs. Serving soldiers posted photos of themselves with rifles, alongside clear racist threats to kill civilians. Young women posted identifiable selfies together with comments like: 'From the bottom of my heart, I wish for Arabs to be torched' or 'Kill Arab children so there won't be a next generation.'[13]

Though whipped up by politicians and religious leaders, this was an online mass movement from below in support of ethnic cleansing – and it quickly moved onto the streets, with demonstrations calling for 'death to Arabs', a spike in arson against Arab homes and attacks on leftist protesters opposing the war.

Far-right racism is of course nothing new, but social networks bring to it two distinct features. First, they create a separate, self-reinforcing mindspace, in which hatred and toxicity become normalized for millions of people. Secondly, the bubble insulates the irrational thoughts and incitements to illegality from challenge, both by political opponents and from independent media sources and human rights groups.

Though the Israeli case is an extreme example, it happened in a country that – like America – is nominally democratic. But there are limits to what an information bubble can achieve for the political right. Though it can shield people from dissenting ideas, it cannot eradicate them. Nor, on its own, can it win an election. Nor does it usually force left-wing and progressive thought into a parallel bubble.

Because of this, in order to achieve power, the modern alliance of 'elite and mob' needed something bigger: a method that leveraged the power of the whole network. During Trump's election campaign, Facebook, Google and Twitter handed them the means to do so: the data-profiling of users and the algorithms designed to serve them targeted content. The content was variously supplied by the real alt-right, plus numerous fake groups and individuals controlled by Russian intelligence. The targeting methodology was supplied by Cambridge Analytica, the data-crunching firm backed by Mercer. 'We did all the research, all the data, all the analytics, all the targeting – we ran all the digital campaign, the television campaign, and our data informed all the strategy,' was how CA's boss Alexander Nix explained the firm's role to undercover reporters in 2018.[14]

There does not have to be collusion, still less conspiracy, for the overlapping networks of right-wing thought-control to utilize each other's assets. Here's how Cambridge Analytica described its operation: 'We collect up to 5,000 data points on over 220 million Americans, and use more than 100 data variables to model target audience groups and predict the behaviour of like-minded people.'[15]

Marketing agencies have tried to do this for decades, of course, but the always-on network gives them a new and massive advantage, both in predicting behaviour and influencing it. Those 5,000 data points might include your regular locations (via GPS or wifi logons); what's in your average shopping basket, who your close friends are, your voting affiliations or your porn-viewing preferences. Once Cambridge Analytica managed to correlate these attributes to voter data, the Trump campaign was able to use the algorithms designed by Facebook, Twitter and Google to target people with advertising to precisely influence voters.

Anyone can buy a targeted ad on Facebook: that's the free market.

Trump spent $150 million on online advertising. But in return Facebook provided him with a team of employees to help his activists learn to use the technology. As communications scholars Daniel Kreiss and Shannon McGregor found in a 2017 study, Google and Twitter offered similar partisan services: 'Representatives at these firms serve as quasi-digital consultants to campaigns, shaping digital strategy, content, and execution.' That is most definitely not the free market: it is corporate collusion with the campaign based on the stigmatization of black people, migrants, disabled people and the media.

With some of it, the overt aim was voter suppression: Facebook helped Trump place content into the newsfeeds of likely Clinton voters that was intended to make them stay at home on polling day. For example, they designed a South Park-style cartoon of Hillary Clinton delivering her infamous 'superpredator' line against black and Hispanic gang members in 1996. It was a racial slur, and Clinton had apologized for it, but now Trump's staff placed the cartoon as a 'dark post' – hidden from everybody except the targeted black voters whose timelines it dropped onto. It read: 'Hillary Thinks African Americans are Super Predators.' A Trump staff member told Bloomberg: 'We know because we've modeled this. It will dramatically affect her ability to turn these people out.'[16]

Facebook, Cambridge Analytica and other social media giants are now embroiled in regulatory investigations. The outrage against these companies has been fuelled not just by the way they targeted advertising, but by the knowledge that Russia was covertly piggybacking onto Facebook's election advertising business in its attempt to manipulate the election. Of 112 campaign groups who bought Facebook advertising on controversial issues in swing states, one in six were traceable to a covert Russian state propaganda agency.[17]

This third mind game – using algorithms to predict and influence voter behaviour – completes the repertoire of control and repression that was developed by the political right in response to the networked revolts of 2011. None of it could have happened without the light-touch regulation culture, and the massive asymmetry of information that comes when single monopolies like Facebook dominate an entire sector.

The stark question facing the progressive majority in advanced countries, as social conflicts intensify, is: could we be defeated not just through censorship, shut-downs and arrests, but by the victory of right-wing logic on the network itself?

The initial answers are not encouraging. The three authoritarian strategies I have cited evolved rapidly and consciously. In response, the behaviour of the protest groups has evolved much more slowly.

The standard operations of a left-wing party or protest group are to mobilize its supporters using networks, but to rely on old, hier-archical institutions – such as the Democratic National Committee or Britain's Labour Party or Syriza in Greece – to gain political power in the old, analogue way.

What we need on top of this is a framework that allows us to take conscious control of our networked 'selves' in ways that prevent them being manipulated, and to fight for a common information system in which the competing claims of political forces can be objectively judged, and the lies publicly categorized as lies.

No term expresses the populist right's fear of freedom more suc-cinctly than the word 'snowflake'. It has a long and varied history as an insult, but its current meaning originates in a speech from the movie *Fight Club* (1996). As a bunch of alienated male skinheads work on their project to destroy consumer capitalism by blowing up Wall Street, an off-screen voice recites: 'Listen up, maggots. You are not special. You are not the beautiful or unique snowflake. You are the same decaying organic matter as everything else. We are the all-singing, all-dancing crap of the world. We are all part of the same compost heap.'[18]

Though scripted as a critique of toxic masculinity in the years when this phenomenon seemed to be on the wane, the movie later became cult viewing among the alt-right.

In 2016, the term 'snowflake' surged into popular use among con-servatives supporting both Trump and Brexit. It came to denote the millennial generation's tendency to take offence at racist and sexist language, to demand 'safe spaces' at universities and 'trigger warn-ings' about material that might upset them; and to shut down the hate speech of racists, homophobes and misogynists. The subtext – that a snowflake quickly melts – was also used to denote weakness.

But actual snowflakes are beautiful. While it is a myth that each one is a unique and perfect hexagon, they do show a very wide range of variation and, in stable conditions, create six-pointed symmetry. In fact, one of the most beautiful things about a snowflake is that it's a physical disproof of magical materialism and the anti-science movement. Its six-pointed structure is completely explained by the inner structure of a water molecule, and by what centuries-old science tells us about the flow of heat.

If you get sick of hearing the proposal that the physical world does not exist, or can't be described by thought, or is only 'created by consciousness' or 'different depending on whether a frog sees it or a human being', think yourself lucky if it is snowing outside. No frog and no idealist philosopher ever gazed at a snowflake and saw seven points. The snowflake is mathematically consistent because reality exists, and because science and rational thought are capable of describing its laws well enough to reduce Kant's unreachable 'thing in itself' to irrelevance.

If you look again at the quote from *Fight Club*, something else becomes clear: the toxic males in the movie are fully signed up antihumanists. Humans, says the protagonist, are part of the decaying crap-heap of organic matter. There is nothing to distinguish us, either as individuals or as a species, from crap. At its very birth, then, the 'snowflake' insult carried the same wider implication on which posthumanism and the new magical materialisms are based. We are no different from plants, animals, stones or faeces.

A further beautiful thing about snowflakes is that they dance. When the wind catches them, if it's cold and dry enough, they swirl in the air. When the composer Claude Debussy wanted to write a piano suite embodying his memories of childhood, he created a soundscape depicting the dancing snow.

For all these reasons I am happy to use the word 'snowflake' as gay activists have used the word 'queer'.

I want to revel in my uniqueness, and in the uniqueness of others. I want to celebrate the difference between a human being like Debussy, who could write *Snow is Dancing*, and a heap of organic matter which could not.

Let the far right, with their conformism, their anti-humanism and their obsession with biological hierarchies own the crap heap. By alighting on the word 'snowflake' they have inadvertently come up with a term much more poetic than 'networked individual' to describe the revolution in self-identity that the technological revolution brings. Let them wave their flags of Kekistan, the swastika, their Nordic power runes and other mystical twaddle borrowed from 'chaos magick'.

If I could design a banner for the movement that will defeat them, it would be a flag with a snowflake – but every example would be randomly generated and unique.

The millennial generation are often criticized for their identity politics, their easily hurt feelings, their detachment from grand narratives and their obsession with defending and curating the small personal space around them. But these qualities can be a source of strength. In fact, when the networked generation chooses to fight, this determination to begin from the self, and defend the self, gives their resistance a hard, granular, irreducible quality.

The networked individual may be oppressed, harassed, crushed down by circumstance. But the life they are living – simultaneously empowered and manipulated by technology – contains the seeds of a project of human freedom based on overcoming this alienation and self-estrangement. The nineteenth- and twentieth-century proletariat was, despite its heroism and self-sacrifice, always designated as a blind agent of change. The snowflake insurrection will be made by people with their eyes wide open.

But to make it, the networked individual has to go through a process similar to that experienced by the working class 200 years ago: the move from atomized survival to the recognition that we have similar interests and a common mission.

For the working class this was done by creating something that after forty years of neoliberalism seems shockingly old-fashioned: a morality. The workers at the Leigh Miners' Gala in the mid-1960s had evolved, over generations, a clear and commonly understood ethical code rooted in the need to make choices other than the ones dictated to them by managers, policemen and politicians.

If this generation wants to defend their right to live a fully rounded human life, to freedom of speech and freedom from surveillance and political manipulation, they need to realize it can no longer be done inside the private and personal world. Because, all over the globe, organized forces are on the march that want to take these freedoms away.

To defeat authoritarian nationalism means taking away its mass support. So we have to organize in a new way. We have to neutralize the political power of the elite and disrupt their mind games with new forms of resistance. We need an economic model to replace neo-liberalism, a new multilateral order that stabilizes globalization and an enforceable global treaty that defends personal freedom.

In the last part of this book I will suggest some actions we can take to get there. But first we need to settle accounts with Marx. The project of radical humanism I propose is based on his biologically universalist theory of human nature. But, given the controversies surrounding it, that has to be placed within a critique of his wider ideas.

PART IV

Marx

Marxism is a theory of liberation or it is nothing.
Raya Dunayevskaya[1]

13

Breaking the Glass

If you want to resist the authoritarian right and fight for basic human rights, you had better get used to being called a Marxist. When the alt-right staged their torchlit parade through Charlottesville in August 2017, their organizer Jason Kessler stigmatized the whole city as Marxist: 'This entire community is a very far left community that has absorbed these cultural Marxist principles advocated in college towns across the country, about blaming white people for everything.'[1]

What is 'cultural Marxism'? Dip into any pro-Trump media brand – from the neofascist websites like Daily Stormer to Fox News – and you will hear a conspiracy theory that, in the 1930s, a group of European left academics called the Frankfurt School brought 'cultural Marxism' to the USA in order to destroy the American way of life. Their supposed weapon was 'political correctness'. Instead of the proletariat, the new gravediggers of capitalism would be women, black people and gays, says the conspiracy theory.

Though it echoes the Nazi term 'cultural Bolshevism', which was used to stigmatize modern art, in its current form 'cultural Marxism' is a term popularized by the conservative American thinker William Lind. Lind argued that political correctness was a form of totalitarian ideology designed to subject white men to the interests of gays, black people and women. It is Lind's understanding of the term – as a plan to undermine the West through promoting social liberalism – that has become a core concept shared by right-wing conservatives, populists and the far right, both in Europe and the USA.

It was to resist 'cultural Marxism' that the neo-Nazi activist Anders Breivik murdered sixty-nine young members of the Norwegian Labour Party in 2011.[2] Breivik's manifesto contains more than a hundred references to cultural Marxism, and a full twenty-seven pages are directly lifted from Lind's work.[3]

By August 2017, the attack on cultural Marxism had reached the White House. Trump adviser Rich Higgins submitted an official memo to the US National Security Council claiming that the president's opponents were operating in a 'battle space prepared, informed and conditioned by cultural Marxist drivers'.[4] Higgins's memo was so off the wall that it even accused the United Nations and the European Union of promoting cultural Marxism. Though Higgins was sacked during the purge that removed Steve Bannon and other civil-warmongers, the attack on cultural Marxism has become a recurrent theme of the racist and misogynist right, in the USA, Europe and beyond.

At one level this is pure paranoia. Women do not demand freedom from sexual harassment because they have read Marx; nor did the black population of Ferguson resist police occupation in Marx's name; nor did the secular youth of Istanbul go on the streets in 2013 toting old books from the Frankfurt School. But in a way the conservative nut-jobs and Nazis have identified what their most dangerous opponent would be, if it existed: a left-wing movement armed with a coherent critique of capitalism, deeply rooted in popular culture, which could link all struggles around race, class, sexuality and gender into a single project of human liberation.

If I could outline such a project without reference to Marx I would do so. Nobody fighting for social justice should have to drag around the stigma that attaches to Marx due to the crimes committed by authoritarian regimes in the twentieth century in his name, let alone what passes in China for Marxism today, which is a mixture of accountancy and Confucianism.

But Marx has to be confronted. As with his contemporaries – from Charles Darwin to Richard Wagner – his impact on the present is so great that he cannot be 'uninstalled' from Western thinking. There is a lot to criticize in Marx's work. But his core idea – that humanity as a species is biologically capable of setting itself free through technological innovation, self-transformation and work – has to form the basis of a twenty-first-century radical humanism.

*

After Marx finished his doctorate in Berlin, he threw himself into the political conflict between liberal republicans and the conservative Prussian monarchy of the early 1840s. His links to left-wing academics who questioned the divinity of Jesus barred him from a teaching position, so he became a journalist on, and then editor of, a short-lived liberal newspaper in Cologne.

Marx was a member of the educated middle class who couldn't make his way in the world; an atheist forced to submit to a religion-obsessed state; a journalist whose every word had to pass through the censorship of a reactionary monarchy. He was obviously going to rebel.

But in fact, Marx conformed for as long as he could to the central project of the German Enlightenment, whose culture he grew up in: that philosophers were doing the important work of humanity, alongside science, by questioning everything and trusting only to reason. By clinging on to the debates within Enlightenment philosophy for so long, Marx carried them into the era of strikes, factories and working-class parties.

By pushing philosophical logic to its limits, Marx fused the two traditions of Enlightenment thinking: materialism and idealism. He learned from the materialists that the world is real, that it exists outside our senses, and that the mind is part of that reality, not separate from it. He learned from Hegel to understand historical change as the product of a long build-up of contradictions inside apparently stable systems, which suddenly break out into major conflicts.

From Hegel's younger followers he adopted radical atheism. By denying God's existence you remove a superfluous jigsaw piece in both philosophy and science. There is nobody coding the great computer of the world before it starts and nobody to press the start button – let alone trying to tweak its outcomes in real time.

Though the philosophers all around him were left-wing materialists, Marx realized that the idealist tradition culminating in Hegel was the only one that possessed a model of change. History, Marx said, is the product of human willpower and imagination. We have the power to choose, but not in circumstances of our choosing.

To the question 'how do I know the outside world exists?' Hegel answered: because you are changing it; because change is a feedback mechanism for your labour and imagination, confirming your ability to alter material reality and to form an accurate picture of it in your brain.

Above all, in the work of Hegel, Marx found a detailed set of concepts to describe the mechanisms through which change happens: via conflict, and through the inner contradictions of a thing bursting through suddenly. Hegel argued that appearances sometimes mask the essential dynamics of a society beneath the surface, and that studying part of a system in isolation was pointless: you had to study the whole thing. This way of thinking, which Hegel called the 'dialectic', proved highly useful as a way of grasping complexity – though as I will argue, it trapped later generations of Marxists in a mental cage.

Sometime between May and September 1843 three new ideas came to Marx.

First, that the struggle against religious superstition is not enough. You have to focus on changing the society that breeds it, and to understand that the impulse to invent gods and then worship them is hard-wired into us as long as there remain gaps in our scientific knowledge.

Second, that the only way to achieve complete human freedom is to abolish private property. When we make something in order to sell it, or buy something made by other people, we are disconnecting ourselves from the most human thing we do, which is to work.

Third, that abolishing property on its own is just the cancellation of something: you could hold all wealth in common but still be trapped by the ideas of ownership; still be disconnected from the product; still – and this is crucial – suffer alienation from yourself and other people.

With these ideas buzzing through his brain Marx arrived in Paris – by now home to tens of thousands of revolutionary-minded, self-educated, violently atheist workers, many of them immersed in experiments to create miniature communist societies. What he wrote during this first contact with the organized working class would unleash the idea that is still terrifying the neo-Nazis, white supremacists and catastrophe junkies who surround Trump. Marx said, simply: communism is the project of individual human freedom.

*

When we ask 'what's the essential attribute of a human being?', Marx says that we should look for qualities that have been constant throughout all the different forms of society created by *Homo sapiens*. For Marx, one such quality is our ability to work to a conscious plan, and in a necessarily social way.

Of course, says Marx, ants, bees and beavers work socially and develop different specializations. After Marx, advances in biology have allowed us to understand that not just animals but plants operate specialization systems, and live in complex ecosystems that resemble a 'division of labour'. Humans are different because they can stand aside from what they are doing and ask the question: 'should I be doing something else?' Marx writes: 'The animal is immediately [at] one with its life activity . . . Man makes his life activity itself the object of his will and of his consciousness. He has conscious life activity.'[5]

But because human production is intensely social, and as far as we know always has been, these biologically given skills – to deploy rationality, to think abstractly and to imagine – also have an intrinsically social dimension. Unlike all other thinkers about human nature, Marx puts labour at the centre of being human.

To explore the implications of this, I want to use the earliest cultural object in existence as an example: the Stadel Lion Man carved around 40,000 years ago, found in a cave in southern Germany in 1939. Shaped from a mammoth tusk, it depicts a human with the head and mane of a lion. In 2009 new shards were studied suggesting it might be female, but for now the figure is catalogued as male.[6]

The Lion Man is very clearly the product of abstract, rational, imaginative, social thinking.[7] It is evidence that from the get-go, humans produced things for other humans. And not just essential things. Modern sculptors estimate that the Lion Man would have taken 400 hours to carve using stone tools – hours that could have been spent hunting or gathering. The smooth patches on its body, from being passed between many human hands, indicate it was probably a ritual object used by a group in the cave where it was found.[8]

This innate tendency to produce for other humans – humans we might never have met, or who may not even be born – is unique to us, says Marx. It constitutes our 'species being'.* But it comes with a down-side.

To understand it, let's look more closely at the Lion Man. It is a superb sculpture: it looks you in the eye; it has the kind of spinal energy a football player or ballet dancer is meant to show. We have no idea what the maker thought about the Lion Man – but the study of surviving hunter-gatherer peoples suggests a range of likely meanings. It could be an attempt to depict a lion in mythical form; or a supernatural being; or the maker's spirit fused with the spirit of a lion; or a human being dressed in a lion's skin – i.e. somebody powerful enough to kill a lion. Or it could have been a storytelling figure akin to a doll or puppet. In all cases, the maker is projecting a human quality into the object, and so are the users.

Marx says, with everything we make, we are externalizing part of our humanity. And while that's true for ordinary tools, it is especially true when we make something with clear symbolic or social use, like the Lion Man.

Marx calls this process 'alienation'. You've probably heard the term 'alienation' used to mean feeling depressed, scared or phobic about work, society and the world in general – and Marx, too, uses it that way. But for Marx, the cause of all the angst is this process of making things for other people, letting go of them, imbuing them with a meaning and with an imaginary power.

The next essential attribute of humanity is language. Animals have language – but only human language is the product of a self-reflective brain. Language, says Marx, is 'practical consciousness' – the ability to present one's ideas to other people and immediately create a shared understanding.

We have no idea what language the Lion Man's creators spoke; we can be certain it was complex enough to tell stories about the other figurines found alongside him. Today, using live brain scans of people skilled at making stone tools, neuroscientists have shown that language and toolmaking use the same part of the brain.[9] During our

---

* It is just in his work upon the objective world, therefore, that man really proves himself to be a 'species being'. This production is his active species life. Through this production, nature appears as *his* work and *his* reality.

evolution into *Homo sapiens*, it is plausible to speculate the one stimulated the other. The important thing about language, for Marx, is that it is another link between our biology and our essentially social nature as technologists.

Marx, like Darwin, had only the basic observational biology of great apes and undated skeleton findings to go on. Today we have a much deeper knowledge of human evolution, both via carbon-dated objects, human cognitive science and the neuroscientific study of our closest relatives, the great apes.

Much of it corresponds with what Marx predicted speculatively: we know that the great apes have some of the thinking skills of human beings, and some basic language. But they do not cooperate. At some point, says Michael Tomasello, one of the leading authorities in early human evolution, changes in our surroundings promoted the survival of groups who collaborated and who began to use language in a way we would now call 'objective' – that is, to describe a world with predictable outcomes. Later, as the collaborative groups began to interact, they normalized the roles needed for the division of labour in a hunter-gatherer clan: using objects and rituals to embody the instructions, they created a culture.

For Tomasello, what differentiates humans from pre-humans is that they 'not only understand others as intentional agents but also put their heads together with others in acts of shared intentionality, including everything from concrete acts of collaborative problem solving to complex cultural institutions'.[10] In the past fifteen years, due to observation of the brains and behaviour of higher primates, he says, we've come to understand culture less as a way of transmitting knowledge for the early humans, more as a way of organizing collaboration. Tomasello's account of early human development helps us understand that, whatever its specific meaning, the social function of an object like the Lion Man was to coordinate human actions around a goal.

In his Paris manuscripts Marx argued that this fundamental trait, of creating tools for social use, is what causes us to then imbue some objects with mythological meaning. Marx calls this 'fetishism'.

Today, you're most likely to hear the word 'fetish' in relation to sex. In Marx's time 'fetishism' was a term used in anthropology to describe the tendency of African religions to imbue objects with a 'spirit life'. With great glee Marx's atheist friends in Berlin pointed out that Christianity itself is a fetish religion: it takes all the characteristics that sum up virtue and projects them into a carving of a dead man hanging on a cross.[11]

Marx, however, wanted not just to criticize religion but to understand how it – and all other kinds of fetishism – arises out of our relationship with the outside world. Here too, everything in modern evolutionary psychology supports the idea that imbuing objects with meaning and with power over us is a fundamental biological trait of humans, intrinsic to the way we developed language. Tomasello believes it is the product of 'group-minded perspective that imagines things from the view of any one of us . . . in the context of a world of social and institutional realities that antedate our own existence and that speak with an authority larger than us'.[12]

Because culture and language are evolutionary products, says Marx, to understand human nature you have to accept it has a history. You have to accept simultaneously that if we could bring the creator of the Lion Man back to life she or he would have the same basic biology as us; but that their social, linguistic and behavioural 'self' would be highly different. While he was still a Marxist, the philosopher Alasdair MacIntyre had great fun at the expense of mainstream social historians over this. He said the only way historians can imagine themselves interacting with the Lion Man's creators is if they, too, were behind the glass case in the museum, mummified and labelled.

Historians, he wrote, can't see a connection between our modern selves and the selves of people of the Neolithic period because 'such a connection could only be established by the concept of a common human nature. And to serve its purpose,' he argued, 'such a concept would have to be historical, have to be a means of showing the past growing into the present.'[13] MacIntyre summed up Marxism as an attempt to break through the glass of the museum: allowing ourselves to understand that human nature includes both biology and history.

If Marx could have seen the Lion Man he might have responded: 'Its creators, just like us, had to work to survive. As soon as they had enough free time they created something that was meant to celebrate that fact – a physically useless but beautiful object which they may have worshipped, or more likely used to project and focus spiritual beliefs, or to tell stories with. It's likely that the powerful gaze and stance of the Lion Man helped reinforce some form of social organization. What I want to know is: what kind of power structures did they use?'

Thanks to the later anthropological study of surviving hunter-gatherer societies we can answer with confidence: their society was egalitarian. Though it would take another 30,000 years to discover agriculture, the people who carved the Lion Man were engaged in their own, very clearly evidenced, technological revolution: the so-called Upper Paleolithic period. The shape of their flints, the combination of wood, stone and bone into complex tools like spear throwers and bows; their ability to hunt more than one species of animal, and to kill large numbers as they migrated; their new burial customs and their cave paintings – all indicate the emergence of more sophisticated societies.[14]

Though there was male dominance in most pre-farming societies (and female dominance in some), anthropologists believe the societies of the Late Stone Age were – so long as they remained nomadic – essentially egalitarian and altruistic. Surviving hunter-gatherer groups are observed to preach 'steadfastly and strongly' in favour of altruistic behaviour, and have been widely seen to ostracize people who try to create hierarchical power. So, though Marx warned us against portraying early human society as Eden-like, evidence has mounted during the past 200 years that such societies did rely on equality and altruism. They welcomed non-family members and distributed basic goods equally in order to survive.[15]

This changed as soon as we invented agriculture. About 10,000 years BCE human societies – though still confined to stone and bone tools – begin to stabilize, domesticate crops and animals, and make their pottery using kilns. Around 3,500 BCE you get the first writing, the first cities and the first smelting of metals. Today we know much more about these early civilizations and their complexities than Marx did. All of it confirms what Marx suggested: with greater complexity and greater wealth comes social hierarchy – or class.

According to Marx, as soon as humans can produce a surplus, a power struggle starts over how it is divided. Those who win it are no longer simply the strongest individuals: they are a specific group who are able to capture the surplus product, and produce a range of justifications for their right to do so.

To get a glimpse of that, let's fast forward to 3,000 BCE and try 'breaking through the glass' with another, very different lion-carving: the so-called Guennol Lioness, produced in the world's first urban culture, Sumer (and currently in the hands of a private collector, location unknown).[16] She is carved from limestone, with a muscular female body and a lion's head. She would have had eyes of lapis lazuli and stood on legs made of gold or silver. As the Sumerians worshipped only gods who took human form, archaeologists assume the lion creature must represent a demon, or even the underworld itself.[17] If so, she was part of the belief system designed to justify the first recorded instance of a class-stratified society.

Mesopotamian cities had a clear class structure: a king; a noble elite who owned land in the countryside and had the right to command the labour and take the produce of the people who lived in the countryside; a lower class who were entitled to own only garden plots, and who had to work in exchange for an allowance of goods provided by the state; and below them the slaves.

This class structure had, in turn, been produced by technological innovation in agriculture and metalwork, which had boosted the output of the land. The first city, in short, was a social system collecting and distributing wealth. And the magnetism that held it all together was religion.

To survey the thousands of artefacts we now possess from ancient Mesopotamia is to see evidence of what Marx called 'alienation' on a vast scale. The lower class is forced by law and custom to alienate its own produce, handing it to the elite as of right. The whole of society then colludes in justifying this arrangement through religion; the fiction that the entire setup has been commanded by the gods. Whatever she is meant to be, the Guennol Lioness is clearly a fetish object – allowing a group of human beings to project their fears, emotions or maybe even just a bond of loyalty, onto a physical thing.

Marx says not only that humans are unique because they make things, but that humans project aspects of themselves into those things, alienating their true humanity, fooling themselves on purpose. As we make history we alter ourselves. But the alteration process is not linear: it produces progress and reversals. The fact that you are probably reading this in a room with electrical light and power indicates that the progress has outpaced the reversals.

Though this seems obvious to us now, the historical character of human nature was a scandal to the most advanced thinkers of Marx's time. Hegel believed history was the unfolding of God's great idea; the materialists believed all of reality was a machine. The whole of nineteenth-century liberalism revolved around the idea of a static, permanent human nature. And this explains why, for several centuries during the rise of capitalism, philosophers had become trapped in a debate about 'determinism' versus 'free will'. Their default position was to believe that everything in history was determined by a previous event, and yet that human beings retained an innate, unchanging capacity for freedom of choice.

Marx's theory of human nature allowed him to tell them: 'History does nothing, it possesses no immense wealth, it wages no battles. It is man, real, living man who does all that, who possesses and fights; "history" is not, as it were, a person apart, using man as a means to achieve its own aims; history is nothing but the activity of man pursuing his aims.'[18]

As a result, for Marx, free will is not something innate and immutable: it is something we possess only partially in societies governed by classes and scarcity. For Marx, free will is something humanity can achieve by changing its social circumstances.

Once we understand human nature to be based on conscious, imaginative labour; once we understand how that labour tends to produce not merely objects but ideals and emotions falsely wrapped up into these objects; once we understand that the structure of all societies has been premised on specific systems of labour and wealth hoarding: then we can see how all advances in technology, productivity and complexity have tended to increase alienation.

So what's the solution to alienation? Marx lived in the most alienated form of society yet invented: the industrial capitalism of the early nineteenth century. Factory workers owned nothing; not their tools, not land, not their time, not even their bodies – which were routinely abused and violated amid a code of silence. The stuff they produced didn't need to be stolen from them: it immediately and intrinsically belonged to their employer from the moment they produced it. Just as today in the sweatshops of the global south, workers were searched on leaving the workplace to deter theft of stuff they had just produced.

In capitalism, instead of the Guennol Lioness, or Christ on a crucifix, the number one fetish object is money. Everything is mediated by it. It seems to have a life of its own – indeed, a power. Money and commodities are what we are obsessed with, in large part because most people don't have enough of either. That is why, from Shakespeare to Molière, the great upsurge of commercial society in the seventeenth century was accompanied by astonishing dramatic depictions of money's power to dissolve all existing bonds, privileges and obligations.

Marx said that to abolish alienation we need to abolish private property. To give humans real freedom of action we need to abolish the power relationships that create a poverty-stricken working class and a wealthy commercial elite alongside each other. You would also have to abolish money and, ultimately, abolish work.

Since none of this could be done without an even more complex social organization than capitalism, and even better technology, Marx understood that the push and pull of history – from ancient Sumer to nineteenth-century Manchester – was the only route to ending human self-estrangement.

This is what 'communism' meant for Marx. But anyone who tells you that communism was his goal is wrong. Marx said abolishing property was only the beginning of human liberation. Once you'd abolished property, you would consciously have to go on fighting to end all forms of self-estrangement, alienation from other people and from nature, and all forms of fetishism – whether religion, money obsession or consumerism. Far from being the 'end of history', said Marx, communism would represent the 'end of the prehistory of humanity'.[19]

For Marx, this was not some lofty ideal or project. It was just the logical outcome of a process we are discovering in much greater detail: our evolution into a species that expresses its shared intent via language and cooperation using technological progress. The more we know about neuroscience and about the evolutionary stages that separated us from other primates, the more Marx's teleological view of human nature looks scientific, not metaphysical.

What makes Marx's concept of human nature relevant today? First, that humanism is under political attack: anti-humanism is core to the alt-right's ideology and anti-humanist ideas have become popular on the left, reducing its ability to resist the right. Second, that the political onslaught in favour of male, white, elite, straight privilege is being prosecuted using technology in an anti-human way. Third, because of the assault on truth.

Truth is only possible if there is verifiable human experience. But now a persistent effort is under way to convince us that truth and rationality don't matter; that we are all partly automata; that we should submit to control by algorithms; that – as Yuval Noah Harari argues – we are 'already algorithms'; that the self is an illusion and that we should let machines think for us.[20]

Marx's theory of human nature is the only one that allows us to confront these attacks and defeat them philosophically. Marx, as we will see, got a lot of other things wrong – but his determination to define humans as something more than the puppets of a great mind or cogs in the machine of history is his greatest legacy to the age of artificial intelligence, quantum computing and genetic engineering.

Because, from the Lion Man of 40,000 BCE until now, we have – despite all the alienation, fetishism and power projection throughout human history – generally maintained meaningful human control over the objects we make. Technological progress leaves most of us unconscious of the way our tools work: few of us could even describe what's inside a smartphone but there is always somebody whose job it is to know.

Information technology has already created new forms of machine control that give its owners vast power. Information technology creates vastly asymmetric access to information between the elite and the rest; and it allows those with the power to impose algorithmic control onto our lives without us knowing, or having the right to know.

With the onset of artificial intelligence we are about to take a step beyond what's been routine for 40,000 years: we will soon be able to create tools that know more than us, and which may quickly develop attributes we cannot control nor even observe. Given our tendency to fetishize things – to imbue brand names and film stars with god-like qualities – it is not impossible that we will begin to see artificial intelligence itself as a cult-like object, and even worship it as the Sumerians did the Lion Goddess. An entire generation swallowed the cult of the market as human controller; there's nothing to suggest we won't swallow machine control just as easily.

Marx's theory of alienation allows us to understand this process and prevent it. It also allows us to understand that, if we want a route to an egalitarian society, towards the complete human-ness Aristotle imagined, we have to go forwards through technological progress, not backwards. In an era when we are likely to see a backlash against artificial intelligence and robotics Marx's theory of human nature remains one of the greatest justifications for technological innovation ever written.

As we stand on the cusp of an era of massive automation, the replacement of human labour by machines and automatic processes on a vast scale, Marx tells us not to hold back in fear, but to seize control.

Faced with the same challenges, almost every other theory of human nature falls apart. If we are merely a 'labouring animal' we won't be doing much labouring in a hundred years' time. If we are just a combination of body and soul, as most religions argue, then to defend the human-ness of the soul it will be logical to retard technological progress.

Liberal individualism, already weakened by decades of fetishizing the market, stands ill equipped to answer the question: 'on what basis do we claim human supremacy over machines once they, too, can develop personalities, emotions and selves?' As a result, it is no surprise to find bestselling authors arguing we have already forfeited the right to control computers.

To build Marx's theory of human nature into a project of liberation through technology, we are going to have to pose the question: 'what did Marx get wrong?' The answer is: quite a lot.

14
What's Left of Marxism?

Just as biological science didn't stop with Darwin, Marxism didn't stop with Marx. Unfortunately, while Marxism in the past hundred years has produced invaluable insights into history and culture, it has also produced abysmal justifications for political repression, crazy economic risk-taking, torture, inhuman social engineering and even counter-revolution. Worse, many of these justifications have begun to resurface among the left movements of the early twenty-first century: among the pro-Assad trolls, the Putin apologists and among older sections of the radical left in Europe, who always secretly regretted the death of the Soviet Union.

To separate what's useful from what's not, let's start with a list of the key propositions of Marx beyond his theory of human nature.

First, Marx assumed that the world is real, material and exists beyond our senses. The problem of how we know this is, for Marx, not solved by a passive description of the relationship of mind to matter, but an active one. When he wrote 'Philosophers have only interpreted the world; the point is to change it,'[1] he didn't just mean that academics need to leave the campus more. He meant that only through the act of transforming the world around us can we actually come to understand it.

Marx tells us that our persistent utopianism – whether in the form of religions or social struggle – is rational. We are not 'destined' to achieve a classless society – there are no gods pushing us around like pieces on a chess board – but if we use the word 'purpose' to mean 'function', then the purpose of humanity is to achieve its own liberation. Aristotle called this a *telos* and, once grounded in better science than the classical world possessed, the Marxist theory of human nature is overtly teleological.

Surrounded by utopian sects who thought they could achieve communism by setting up communes and sharing out their goods, Marx insisted that the route to the future society, in fact, lay through capitalism. It is the first economic system compelled to revolutionize productivity and continually blow away fixed hierarchical structures. This, Marx said, creates the conditions for replacing capitalism with something better. Because capitalism was the most acute and final form of class society, he assumed that once you get rid of it all forms of class hierarchy and inequality should disappear along with it.

Society is not just a mass of individuals. The history of all previous societies, says Marx, is the history of class struggles. Sometimes the struggle leads to the victory of one class over another, and a new form of exploitation begins, as when the French bourgeoisie overthrew the aristocracy in 1789. Sometimes, he says, it just ends in ruination – as with the Roman Empire, swept away by tribal invasions from the East and the inefficiencies of its slave system.

But Marx understood that class struggles are fought using ideologies like religious or mythological beliefs, which can mask the fact that historical events are driven by economics and power. For example, the Conquistadores told themselves they were going to Mexico to convert its inhabitants to Christianity on behalf of the divinely ordained king of Spain. In fact they were going to kill the indigenous people, steal their gold and, in the process, fuel the rise of merchant capitalism in Europe.

Capitalism, said Marx, is the last and most advanced form of class society. But it leaves humanity – above all the working class, which forms the majority – at the maximum point of alienation. Unlike all previous subordinate classes, said Marx, the working class owns nothing. If the material interest of a landowner is rent, and that of a factory owner profit, the true material interest of a worker is to overthrow them both: to abolish private property and replace it with a regime of common ownership.

Marx called the opening acts of a socialist government the 'dictatorship of the proletariat'. By this he meant that, as in the *dictatura* of ancient Rome, the working class, if it gained power, would have to impose temporary martial law in order to suppress the resistance of the rich and powerful.[2] He based this assumption on the fact that in his lifetime every attempt by workers to achieve things democratically, or to push democratic revolts towards social reform, had led to them being massacred by the bourgeoisie.

But how could the working class become revolutionary? From his first days in the Paris of 1843 to his death in London forty years later, Marx placed himself in the company of radicalized workers, wrote about their conditions, listened to their ideas and gave them advice, usually while drinking large amounts of alcohol. He knew that while the mass of working people were trapped by lack of education, some had been able to break out of the ideologies imposed on them, both consciously – by bosses, aristocrats and priests – and subconsciously by the power relationships in the factories they worked in.

Just as human beings in general achieve knowledge by interacting with the world, the working class clears the ideological fog in its head by engaging in collective struggle. Marx called this 'the alteration of humans on a mass scale' and believed it could take place only during a revolution.

Theoretically, however, he believed the working class could accomplish its historic mission independently of what was in its head. He wrote: 'It is not a question of what this or that proletarian, or even the whole proletariat, regards as its aim. It is a question of what the proletariat is, and what, in accordance with this being, it will historically be compelled to do.'[3]

One event changed Marx's understanding of the process of revolution: the Paris Commune of 1871. During this extraordinary uprising, after the French army abandoned the city, the working class exercised control at many levels, from the official 'commune' or city council, through to revolutionary assemblies, clubs, women's groups, trade unions and co-ops. The Commune was in many ways a semi-state and after witnessing it (he was in constant contact with sympathetic activists during the uprising) Marx became convinced that, when they took power, the working class would abolish and dissolve a state bureaucracy and parliamentary institutions standing above the people, ruling instead 'from below'.

In *Capital*, Marx shows how workers are exploited by their employers coercing and extorting extra work from them, above the value of the product they're producing. This, for Marx, is the only source of what today's businesses call 'value-added'. Capitalism can survive and renew itself for a long time, says Marx, because growth and productivity produce rewards for the worker as well as the capitalist.

But the process of accumulation – investing, making profits, banking them or reinvesting them in new machinery, skills, products – creates spontaneous breakdowns: in the form of commercial crises, where supply exceeds demand; in the form of banking crises when credit dries up; in mismatches between the consumer economy and heavy industry; and ultimately in the long-term exhaustion of previous innovations, which place downward pressure on the rate of profit.

Because everything capital does is designed to replace work with machines, says Marx, it must constantly create new demand: new higher-value products, with higher wages for the workers who produce them, so that the population has the spending power to maintain demand.

In a notebook known as the *Fragment on Machines*, Marx imagined how this clash between technology and social structures within capitalism might play out if we developed to the point where all technology was reliant on 'social knowledge' – that is to say, on commonly understood techniques, definitions, instructions, workflow patterns and so on, rather than on people operating simple machines. Once knowledge is social, said Marx, production cannot be privately controlled. If we ever manage to embody social knowledge in a general information store – a 'general intellect' – it will, he said, blow capitalism sky high.

For Marx, then, there is both a material and a knowledge dimension to freedom. He defines communism as 'free men, working with the means of production held in common, and expending their many different forms of labour-power in full self-awareness as one single social labour-force'.[4] The forces we've unleashed through technological progress stop working 'behind our backs': we take control.

This is not the whole of Marx – but it's the essence of his thought.*
Now I want to interrogate these ideas using everything we've learned
since Marx formulated them. I am not interested in the standard
attacks you hear from those who believe markets are the ultimate form
of rationality, or that power inequalities are natural, or that Marx was
an asshole because he got his housemaid pregnant. What I'm interested
in is a Marxist critique of Marx around the issues that confront us
today: women's oppression, climate change, how to understand complexity, how to abolish scarcity and how to impose human control over
thinking machines via a global ethical framework.

The standout omission from classic Marxism is an account of wom-
en's oppression and the domestic labour that helps sustain capitalism.
Marx understood that women's oppression and exploitation were
essential to the power structures of all class societies. But his account
of how women's exploitation props up the entire system was inadequate and therefore wrong.

Marx said the value of a worker's wage reflects all the inputs
needed to present himself, or herself, as a labourer at the factory gate:
that included all the bread and all the baking done in the commercial
sector, all the tailoring and all the schooling done outside the home.
But Marx never included in the calculation all the sewing, darning,
cooking, washing and child-rearing done inside the home itself by
women.

He regarded the patriarchal working-class family – with a male,
skilled worker going to work and the wife and family surviving on his
wage – as a doomed institution. It was logical to assume capitalism
would put all women and children to work because that was what the
capitalists continually said they wanted: female and child labour
working hours so long there was minimal time for domestic labour.

So Marx paid scant attention to the way women's unpaid labour in
the home, and their function as birth mothers and child-rearers,

---

* Also it's not 'Marx & Engels' – Friedrich Engels was Marx's collaborator but took
a different path towards communism initially, and after Marx's death – scholars
now believe – built some of his own early ideas into what became the official version
of Marxism.

contributes to the wealth and power of the elite. He failed, in short, to understand reproductive labour as a specific form of exploitation vital to capitalism.

In the 1960s, instead of waiting for communism to bring about the end of their oppression, a generation of women decided to start fighting for liberation directly. Silvia Federici, a key thinker in Marxist feminism, describes the new strategy as a mass 'refusal of work'. Surveying the breakdown of family values, Federici writes: 'The collapse of the birth rate and increase in the number of divorces could be read as instances of resistance to the capitalist discipline of work. The personal became political, and capital and the state were found to have subsumed our lives and reproduction down to the bedroom.'[5]

As a result, argues Federici, neoliberalism had to change the way reproductive labour supports the profit-making process. After the Second World War, those countries that encouraged women into the workforce did so through the state provision of childcare, elderly care, communal laundries and family welfare payments. Neoliberalism, from the late 1970s, privatized and commercialized reproductive labour.

In the Doris Day era, the working-class family had been a machine for producing a male breadwinner through unpaid domestic work. Now it would be a consumer of commercially provided services on a vast scale. Women were drawn en masse into the waged workforce – yet in addition they were still required by culture and tradition to do unpaid domestic work, above all rearing children in their early years.

This has intensified the battle over men's assumed 'right' to higher wages, to job seniority, to sleep around while stigmatizing women who do the same, to perpetrate domestic violence and to do things to women's bodies without consent.

I have described above how life under neoliberalism became 'performative': obey the ritual required by market interactions and you will survive. As the free-market performance becomes meaningless, we are beginning to understand how this performativity allowed a deep-seated misogyny to survive in private and online spaces. You can bet that many of the same losers churning out misogynist bile on the 'chans' work for corporations where they are routinely required to affirm their support for equal opportunities and decry sexism.

While twenty-first-century misogyny is a new iteration of an age-old theme, we nevertheless need to be aware of its technological and situational novelty. This is the first time a model of capitalism has broken down without an elite reaching for an alternative model; it is the first time a model of capitalism has broken down while tens of millions of women have been experiencing economic, sexual and behavioural freedom. It is a dangerous moment – and to resist the woman-hatred being spread by the authoritarian right, we need a lot more than Marx's theory and the traditional tactics that theory has inspired.

Both Marxism and feminism include biological claims. Marx's theory of human nature is not gender specific: it says we are all defined by imaginative, goal-centred work and that, once we overcome scarcity, all forms of power hierarchies should disappear. Feminism says both male power and female oppression can be biologically determined: there has to be a parallel struggle, with separate dynamics, and it will have to carry on beyond the achievement of what Marx called communism.

Given that every form of capitalism, every workers' state and every progressive movement has reproduced women's oppression, the evidence supports thinkers like Federici, not Marx.

The second big inconsistency in Marx concerns the working class and its role in history. Sometimes they are destined to perform the conscious overthrow of capitalism; sometimes they are its unconscious gravedigger.

The decisive mistake he made about the working class can be deduced from the German word he used to designate their historic role: *träger*, which means bearer. For Marx, the working class were the bearers of an implicit need to attack private property and destroy class domination, and at the same time bearers of the fate of capitalism: its gravediggers.

Throughout the entire history of the industrial working class, this proved false. For the past 200 years the workers' movement has been the most heroic and consistent force in fighting for democracy, social progress, internationalism and women's rights. But at no stage did the majority of working-class people consistently and effectively support a project of abolishing private property.

Instead of embodying (or 'bearing') the antidote to private property, workers actually embodied their own interests as a class within capitalism: they demanded higher wages, equal rights and a higher social wage. When their struggles went beyond this – as they frequently did – they often settled for control instead of power, above all control at work and the right to live an autonomous cultural life.

Only when pushed to the limits of toleration by dictatorships, chaos, the military defeat of countries or by fascism – as in Paris in 1871, Russia in 1917, Bavaria and Hungary in 1919 or Spain in 1936 – did something like a majority of workers opt for revolution. Even then, as a class they proved consistently unable to keep control of political power, being quickly usurped by privileged groups from within the revolution itself. In both the Russian and Chinese revolutions, once a bureaucracy had seized control, the working class settled for a version of what they'd originally asked for – an element of control over production – and something they had not: a privileged position compared to the peasantry.

Today, though more than half the adults on the planet work for wages, the culture and solidarity of the old working class has been eroded. Those who dreamed that it would be revived in the new industrial heartlands of China and Latin America were only half right. As I reported in my book *Live Working or Die Fighting*, class struggles and self-organization among this new working class are rife – but so too are the neoliberal self and networked individualism, often combined with the cultural hangovers of peasant life, such as village networks, mafias and nationalist illusions. As a result the modern, global working class no longer thinks or acts like the classic proletariat of the twentieth century – and no amount of exposure to the class struggle will remedy this.[6]

We can understand why Marx got it wrong. All around him were working-class people who owned nothing and who were deeply alienated from capitalism, the Church and even traditional family structures. Working-class militants believed their only option was to abolish private property, just to free themselves from its imprisoning demands. In them Marx found a social force that fitted perfectly into Hegel's theory of history. Here was the living contradiction of capitalism, negativity made flesh, the bearer of the new society. Its historic purpose was to overthrow 10,000 years of social hierarchy.

Once realism set in, both halves of the twentieth-century left were shaped by the tacit admission that it was not happening as predicted. Leninism was premised on the idea that on its own the working class could achieve only 'trade union consciousness'. To trigger the revolution, an elite of intellectuals and educated workers formed into a hierarchical party was needed. Mao went even further, suggesting that the experience of the urban working class – which in Shanghai and Guangdong had staged massive but doomed revolutionary uprisings in the 1920s – meant that the peasantry was the true revolutionary class.

Meanwhile, the social democratic wing of the left concluded that the immaturity of the working class, their lack of culture and education meant that a long period of parliamentary activity was needed – and in a mirror image of Leninism, they insisted this, too, would have to be led by lawyers, intellectuals and professional politicians from the middle class.

In fact, the most politically conscious workers repeatedly defied both Lenin and the reformist moderates. They struggled for far more than just wages and trade unions but much less than socialist revolution. The leitmotiv of working-class history, occurring again and again, is the creation of islands of control and freedom within capitalism. One thing that the Leninists, Maoists and moderate social democrats agreed on was that such islands of self-control were a distraction.

A second interesting thing the working class did, again in defiance of the Marxism of their leaders, was to create an alternative morality.

Marx's critique of capitalist exploitation is full of morally charged rhetoric. But he spurned the idea of a left-wing moral philosophy. Whenever he heard the words 'moral philosophy' he was said to be in the habit of laughing out loud. Likewise, most nineteenth-century socialists despised moralism, both the kind emanating from the Church and that from the liberal do-gooders, who constantly preached to workers to accept their lot rather than strive for something better.

Marxism was a philosophical revolt against the claims of

Enlightenment thinkers, like Kant, to have discovered an eternal morality that existed independent of humanity's social evolution. Whose morals, asks Marx, does a woman follow when, forced by poverty, she sells sex to the factory owner? Whose morals are the devout Christians of the American slaveocracy following as they whip fellow human beings to death? All moral systems, Marx believed, are a reflection of the class hierarchies that produced them.[7]

If so, Marxism's failure to produce a moral or ethical code stands in sharp contradiction to the actions of the working class, which continually tried to do so. In fact, the whole process of becoming a class 'for itself' – which, Marx said, the working class must do to achieve socialism – was in practice a moral project, based on exactly the kind of relationship between ends and means implied in Aristotle.

Though at times hypocritical and always patriarchal, working-class morality understood that there had to be something more than 'the end justifies the means'.

When labour movement cultures were strong, there was a conscious attempt to build a community in which all acts contributing to the good life were seen as virtuous, and in which living an exemplary life and possessing virtues – such as solidarity, generosity and the capacity for self-sacrifice – was as important as the 'end' itself (whether it be winning a strike or overthrowing a government). Since we want to survive within capitalism, educate ourselves and expand our control within the workplace, workers told each other, this is how we have to behave.

Richard Llewellyn's sentimental but highly realistic novel *How Green Was My Valley*, set among Welsh miners in the early twentieth century, captures that morality perfectly. There is a fire-breathing evangelical preacher, there is a police force, but the moral codes the miners follow are independent, complex and unwritten, always designed to hold them together as a working community and prevent them competing with each other.

Because they refused to express a moral system of their own, Alasdair MacIntyre once complained, whenever Marxists were confronted with an ethical dilemma they became either Utilitarians or Kantians. They either said, as Trotsky did in his 1938 essay *Their Morals and Ours*, 'the end justifies the means';[8] or they proclaimed 'eternal' moral principles which were usually just pale reflections of the Christian commandments.

If that is so, then the real practitioners of virtue ethics were the working class. It was they who evolved new norms of behaviour and categories of right and wrong from an understanding of their own destiny within a given community.

But when, from the 1960s onwards, Marxists began to face the problem of working-class culture's decline and atomization, they did so via expressions of despair. Herbert Marcuse, one of the 'cultural Marxists' with whom the alt-right are today obsessed, believed the industrial proletariat had become one-dimensional, bought off by consumerism and sexual promiscuity, and that the role of fighting for the future had fallen to the oppressed groups: women, minorities and people in colonial countries resisting imperialism.

André Gorz, a French Marxist writing in the 1980s, went further. With no 'bearer' to fulfil a historical role modelled on what Marx read in Hegel, communism was just another utopia. But, he said, we should go on fighting for it anyway, without the comfort blanket of historical inevitability.[9]

I see the situation differently. Having destroyed and dispersed the industrial proletariat, neoliberal capitalism has reincarnated its gravedigger in a new form: the networked individual. The networked individual 'bears' the characteristics of future liberated humanity much more clearly than the coal miners of my grandfather's generation. If they do overthrow capitalism, networked individuals will do it consciously and gradually, not as the unconscious puppets of historical forces. And they have the collective interest to do so, for the following reasons.

First, information technology creates the opportunity to build islands of abundance and self-control inside capitalism, bypassing the stages of scarcity, planning, rationing and centralized control. In the early twenty-first century numerous left-wing thinkers, including myself, had the same thought at once: that information technology, by collapsing the price mechanism and enabling rapid automation, makes it possible to aim straight for the goal of a classless, cooperative and fully automated society. So the networked individual has an achievable goal.

Second, s/he has an existential reason to resist. The crisis of neo-liberalism can be solved only if it starts pushing market relationships even more coercively into the lives of everyone beyond the elite: invading our bodily existence with control, commercializing our lives, collecting data on our every movement, nudging and controlling our behaviours via algorithm, forcing competition into areas in which we currently collaborate. In the twenty-first century, if capitalism survives, it will do so by forcing the majority of us to exhibit the qualities Foucault observed: to remain 'eminently governable' and to compete with each other viciously as 'entrepreneurs of the self'.

Third, the networked individual is a cog in the wheel of capitalism in a much more complex way than was the industrial worker. Networked individuals exchange labour for salaries in the old way. But in addition their borrowing and saving is all that sustains the finance system. On top of that, they are increasingly 'prosumers' – their acts of consumption create brands, and their acts of choosing and sharing knowledge are what build the vast data stacks on which the market valuations of Google, Amazon, Alibaba and so on are based. Capitalism has become, as the Italian Marxist Mario Tronti put it, a social factory. Streams of profit flow into the capitalist's bank account from our activities both at work and beyond work.[10]

As a result, if we were to reject the norms, routines and performative culture of neoliberalism we could create big trouble. Forms of resistance that were seen by classic Marxists in the 1960s as merely 'cultural' – such as the consumer boycott, the brand damage campaign or forming co-ops – can today be economically, materially and systemically harmful to capitalism.

Finally, as neoliberalism falls apart into competing power blocs, triggering more extreme forms of authoritarianism and the rise of the alt-right, all our current limited manifestations of freedom will be attacked. If you want a vision of the future, to rephrase Orwell, picture an army of trolls and bots working for a kleptocratic president, threatening to rape a female journalist whose address they just published on the internet.

In the face of the evidence, clinging to the Marxist theory of the proletariat goes against the spirit of Marxism. But if I am right, and the networked individual is the agent of the next big change in history, then we have to do in the 'social factory' what our grand-fathers did in the industrial factory: find each other and act. In doing so, the revival of a collective plebeian moral practice is one of the most important challenges facing us today.

In 1859, Marx outlined a general theory of how modes of production rise and fall. So long as the economic structure is promoting techno-logical progress, it survives. When it becomes a 'fetter' for progress, it falls. This process involves an interplay between the economic struc-ture and the cultural, legal, social and ideological superstructure.

Yet nowhere in the three massive volumes of Marx's later master-work *Capital* is there any concrete prediction as to how this might happen. There is a theory of value, which stands the test of time and actually explains better than mainstream economics the disruptive effects of information technology.[11] There is a theory of crisis, whereby technological innovation replaces labour with machinery, forcing the superstructure of capitalism constantly to mutate: to cre-ate new needs, to create higher-skilled jobs, new work patterns, new hierarchies in the workplace – even, Marx argues, forcing developed countries to colonize poorer ones and export their surplus pop-ulations there.

However, Marx could not anticipate the large adaptive mutations in the economic structure of capitalism itself which happened after his death. He described one such mutation: the survival of industrial cap-italism after 1848, by making a strategic truce with the working class and inventing the stock market system. But he never theorized it.

On 200 years of evidence, you might revise Marx's 1859 summary as: 'if the economic structure becomes a fetter on the development of technology, it usually goes through a traumatic mutation in order for capitalism to survive'. But that would leave you with a theory of cap-italism's survival, not its demise.

However, in *The Fragment on Machines* – a document written in 1858, just before the famous 1859 *Preface*, and with the same thought process informing it – Marx does predict how capitalism's clash with technological progress could end up destroying the whole basis of a market-driven economy.

The preconditions for this are as follows: (a) that machines have substantially pushed labour out of the production process; (b) that technological progress is taking place at the level of information rather than physical activity (i.e. the design of machines, the automatic functioning of machines, the redesign of workflow processes); and (c) that such progress relies on the socialization of knowledge. Once the work and knowledge of everybody are contributing to the productivity and efficiency of everybody else, via what Marx calls the 'general intellect', an absolute contradiction appears between technology and private property: the ultimate social form of capitalism.

In this scenario. the economic system has developed to a stage where it is using science and technology to make the creation of wealth depend as little as possible on work. Yet science and technology contain implicitly social knowledge, at odds with the economic structure based on private companies and intellectual property. Any technology based on socialized knowledge, Marx predicted, blows the foundations of private property 'sky high'.[12]

A huge theoretical war has been waged by orthodox Marxists against *The Fragment on Machines*. It represents the unwelcome intrusion of a humanist vision of technological liberation into an ideology of class struggle.[13] Yet sketchy as it may be, the *Fragment* contains what *Capital* does not – a concretization specific to info-capitalism of the prediction Marx outlined in the 1859 *Preface* about how economic systems fail and die.

Marx wrote: 'No social order is ever destroyed before all the productive forces for which it is sufficient have been developed ... Mankind thus inevitably sets itself only such tasks as it is able to solve.'[14]

This open, non-specific prediction serves as the best guide to the dynamics unleashed by the clash between information technology and the economic structures of markets, waged work, intellectual property and algorithmic control. A complete materialist theory of how capitalism ends will probably be written only after it has happened.

If you made a list of all the claims that have been made about cause and effect since the scientific revolution began, it would run chronologically from simple to complex. In 1611, Kepler, noticing that all snowflakes have six corners, speculated that, as it freezes, 'the smallest natural unit of a liquid like water' probably crystallizes most efficiently as a hexagon.[15] In the nineteenth century, when they were able to understand atoms and molecules, scientists offered a more complex explanation of a snowflake. Today we can also understand a snowflake as a 'fractal'. Chaos theory tells us to see a snowstorm as a complex system which has become unstable. When a grain of ice forms, and begins to draw the instability in the system towards it, it forms the tiny branches of ice we call a snowflake.

Kepler understood a snowflake as a thing; modern physics understands it as a linear but reciprocal process; chaos theory understands it as a non-linear process involving unpredictable feedback loops between two systems, the water molecule and the snowstorm.

This timeline – from simple one-way explanations to relational ones, to complex, chaotic and uncertain ones – also happened in the study of society. The problem for us is that the method Marx used to describe complexity is not good enough. It's called the 'dialectic'. It was better than the simple, one way, cause-and-effect explanations it replaced and may still be a useful framing device. But, used to the exclusion of other methods of analysis and explanation, it drove Marxism off a theoretical cliff.

Let's start by ditching the notion of 'dialectical materialism'. Marx never used the term, but after his death Friedrich Engels tried to codify dialectics into a theory of everything. Posing as a complete scientific theory of reality, 'diamat' was taught to millions in Stalin's Russia and is today being revived by Xi Jin Ping as political cover for his power grab.

The core mistake Engels made lies in the claim that Hegel's law of development through contradiction, or dialectic, was 'an extremely general law' which 'holds good in the animal and plant kingdom, in geology, in mathematics, in history, and in philosophy'. It is true that apparently dialectical processes can be observed in nature. Humid air becomes snow – that's a dialectical transformation, for example.

However, if we insist, as Marx did, that our mental model of the world has to be derived from our best scientific understanding of reality, it would be ludicrous to suggest that the dialectic is the finished form of that model. It would be like declaring that the history of music stops with Beethoven.

We can say, as Engels did in his *Dialectics of Nature*, that 'the whole of nature, from the smallest element to the greatest, from grains of sand to suns, from protista [amoebae] to men, has its existence in eternal coming into being and passing away'. But that does not mean a set of logical propositions developed in the late eighteenth century is capable of accurately describing that process. And indeed, Engels's non-accidental use of the word 'eternal' – that's to say his implicit refusal to countenance the heat death of the universe – looks to us today deeply un-materialist.

The first big clash between Marxism and science – Lenin's attack on Ernst Mach, the man who discovered shock waves – stimulated the first coherent rethink about dialectics from within Marxism. In response to Lenin, the Russian physician and Bolshevik activist Aleksander Bogdanov warned that, by sticking to dialectics as a dogma, Marxists risked allowing logical categories to obscure the dynamics of reality.[16] This, in fact, had already begun: almost every mistake made by Lenin and his followers originated from trying to fit complex reality into a simple scheme, and from the conviction that capitalism could not recover its dynamism because dialectics stipulated its imminent demise.

In the Paris manuscripts of 1844 Marx wrote that, once human beings had reconnected with nature and abolished both private property and the state, 'natural science will in time incorporate into itself the science of man, just as the science of man will incorporate into itself the science of nature: there will be one science'.[17]

Given that was his hope, Marxism as a social science must therefore learn from natural science, must adapt and utilize its mental models of complexity, chaos and uncertainty. Just as the dialectic was an attempt to deepen and 'think beyond' eighteenth-century rationalism, so too must a twenty-first-century left be prepared to think beyond the dialectic, and to draw on logical frameworks arising from scientific observation.

How do we do that? We continually look again at the evidence and go on asking questions.

The last thing Marx got wrong, and in some ways the most important, concerns the ecosystem of the earth. At the theoretical level,

Marx understood that humanity is part of nature, but a unique part: we can transform the natural world towards human-centred goals. He insisted that human societies are 'not owners of the earth. They are simply its possessors, its beneficiaries, and have to bequeath it in an improved state to succeeding generations'. His collaborator Engels warned that we should not 'flatter ourselves overmuch on account of our human conquest over nature. For each such conquest takes its revenge on us . . .'[18]

Though Marx knew nothing of climate science, he did at least theorize the inevitability of the two systems – human society and nature – coming into conflict once you reach a mode of production based on the unlimited search for growth and productivity. Urbanization and the commercialization of agriculture, said Marx, were destroying the two sources of all wealth: the soil (through nutrient depletion) and the human being (through falling life expectancy, poverty, ignorance and epidemics).[19]

Nevertheless, in his response to Thomas Malthus, who claimed that capitalism would be destroyed by overpopulation, Marx saw no hard environmental limits to capitalism. Though he occasionally accepted such limits might exist, he assumed that technological progress could outpace the natural limitations imposed by raw material scarcity and the exhaustion of land fertility. The idea that fossil-based energy systems were going to destroy the earth was as absent from his thinking as it was from everybody else's at the time.

But that does not absolve Marx. The assumption behind his critique of Malthus was that there were no natural limits to the expansion of capitalism – only an inevitable clash between highly productive technology and the old social forms of class and private property. His view of capitalism was essentially optimistic: technological progress can always solve the problems it creates. And that had practical consequences: it encouraged Soviet Russia to pollute and consume the natural environment to the point of destruction. And until very recently, it authorized the Chinese bureaucracy to do the same.

Today, as the result of more than 200 years of industrial development, we can reframe the problem. Man-made climate change does represent an absolute limit to capitalism. Climate science predicts that if the earth warms beyond two degrees above its long-term average, chaotic feedback loops will emerge within nature itself, accompanied by socially catastrophic natural events.

As they hit this new problem, contemporary Marxists came up with three distinct approaches. The first was to continue with the techno-optimism of the nineteenth century: to search for a technical fix that would reverse climate change, while arguing that capitalism has to be overthrown to achieve it.

A second trend, associated with the left-wing US economist James O'Connor, argued that Marxism has to be expanded into an account of 'two contradictions'. The first is the familiar one, between technology and the economic structures surrounding it. The second, O'Connor wrote, is between capitalism and the commercialized natural environment it has created: tilled fields, polluted air, the global system that grows legumes in Kenya and flies them to Britain on a jumbo jet – and ultimately the atmosphere's ability to absorb carbon.

The strength of O'Connor's thesis is that it builds on an insight into capitalism that later Marxists understood, but which Marx himself did not: capitalism's need, as a system, to constantly interact with and consume other systems. To O'Connor, climate change signifies the limit of capitalism's ability to go on transforming nature.

However, full-blown eco-Marxists believe that even O'Connor underestimated the scale of the problem: that he was trying to provide a 'Marxist' rationale for limiting carbon use when in fact there is a purely ecological case for limiting it – and for radical action to address numerous other critical threats to the ecosphere.

Faced with the fact of man-made climate change, and the sophisticated attempts by twenty-first century Marxists to understand its social implications, it's clear the writings of Marx himself are inadequate. But if we follow Marx's advice to 'analyse the whole thing', the whole thing has to be the earth's biosphere, the human population within it, and their current technologies and social structures.

All of which has obvious implications for any project to use technology to move beyond capitalism: it means consciously managing the interaction between economic development and the climate. It means, in practical terms, ending carbon use and creating a 'circular economy' which allows us to reduce raw material extraction massively, in a way not anticipated by Marx's one-directional optimism about productivity and growth.

Marx's thought contains major gaps, mistakes, non sequiturs and false turns. And not over side issues, but some of the biggest problems that confront us. Given all that, why are the populist right so terrified of Marxism? The answer is not simply because in the hands of a few German émigrés it created the rationale for political correctness. It is because, stripped of its authoritarian impulses, it can still be the most important source of a radical strategy of resistance.

In the 1950s, in parallel with the despairing social commentaries of the thinkers currently targeted by the alt-right – Marcuse, Adorno and Horkheimer – another current developed, one of more relevance to us today: the Marxist humanism of thinkers like Raya Dunayevskaya, the Chicago labour organizer who first translated the 1844 manuscripts; and the self-proclaimed 'radical humanism' of figures like Erich Fromm.

One of the major figures in humanist Marxism, the historian Edward Thompson, declared after the Hungarian uprising of 1956 that 'I can no longer speak of a single, common Marxist tradition. There are two traditions.'[20] There was a tradition of humanist Marxism dedicated to freedom and an anti-humanist one dedicated to the justification of oppression, and which reduced the agency of human beings. The anti-humanist tradition, he wrote, had to be fought to the death. Thompson said that, if that Marxism meant the subjection of human beings to historical forces and the eradication of their power to change the world, he would rather convert to Christianity or even plain liberal moralism.

For us today, the main purveyors of anti-humanism and fatalism in left-wing garb are no longer the pipe-smoking communists of Althusser's era: they are the postmodernists, 'object oriented ontologists' and post-humanists who have, thankfully, forgone all claims to Marx's legacy. So there is no need for anyone who wants to defend Marx's humanistic principles to reach for the rosary beads. But we do have to acknowledge – and be proud of – the continuity of our ideas with the human-centred religions of the Axial Age, and with the Judeo-Christianity of the Enlightenment.

So here's how I would answer the question: 'are you a Marxist?'

I am a radical humanist who thinks we're on the cusp of achieving something Marx wanted: a technologically enabled society in which most things are consumed for free, and the alteration of human beings on a mass scale in order to take advantage of such freedom. Like Marx, I believe this propensity to achieve freedom is the product of our evolution, and every recent advance in genetics, evolutionary biology and neuroscience reinforces that belief. Like Marx, I believe the socialization of knowledge through technological progress will bring us up against the limits of a society based on private property.

But unlike Marx, I believe this human revolution will be achieved not by the blind actions of a single class, but by a diffuse network of human beings acting consciously. Unlike Marx, I believe the planet creates limits to the way humans should use technology and mandates certain priorities in the transition beyond capitalism. And unlike Marx, I don't laugh out loud at the words 'moral philosophy' – because the nature of the technology we will rely on to achieve abundance means we need a global ethical framework to keep it under our control.

PART V

Some Reflexes

In a class divided society human possibility is never fully revealed ... and because of this human development takes place in quite unpredictable leaps. We never perhaps know how near we are to the next step forward.

Alasdair MacIntyre[1]

Interlude . . .

Suppose there was a planet containing millions of species, out of which maybe a handful achieved – by complete accident – the ability to think consciously, make rational decisions, use language and develop tools.

Their technological history might move slowly: for about 3 million years they make only basic stone tools. Then, one of the species develops a culture, a more complex and varied social structure and a richer language, which helps it to spread geographically across the planet while the other thinking species die out. This process takes, maybe another 300,000 years.

Then things start moving faster: from the first cultural object to the first settled agriculture takes maybe 30,000–40,000 years. From pottery to bronze, from bronze to iron, from tribal chiefs to an explicit theory of democracy, plus a bunch of religions promising the future self-realization of this species, takes maybe 10,000 years. About 2,000 years later you get steam engines. Within a hundred years of their widespread deployment, the productivity of this species – having flatlined since the first cities and agriculture appeared – takes off at a 45-degree angle. Then, as machines with moving parts give way to digital machines, material productivity in some sectors goes exponential.

What is the most likely mindset among the members of that species lucky enough to be alive at this amazing moment of take-off? Surely it must be euphoria, confidence, the conviction that – for all the problems they are beset with – further progress is possible?

As we have seen, it is not. The dominant mindset on our planet is fatalism. The dominant political ideology is worship of the market.

After revelling in the 'end of history', many liberal and educated people are in mourning for the fact that history has returned: in the present chaos they can see only the threat that history will rewind towards fascism and dictatorship, and that the damage our species has done to the planet will be made worse and become irreparable.

Now imagine that some members of that species wanted to snap out of such fatalistic thinking. To do so they would need to decide to take a combined series of actions: to adopt a different economic model, to revive more diverse and resilient forms of democracy; to uphold the universality of human rights, and to launch grassroots collaborative projects to rebuild solidarity between human beings.

To pursue these projects, suppose large numbers of educated members of this species decided to do what the working class of the nineteenth century did: to find each other and act. Suppose they tried to move from understanding their common interests – what Marx called becoming a 'class in-itself' – to becoming 'for themselves', and fighting for a positive goal.

If so, they would have to go beyond a bullet-point list of policies and demands on their governments. They would have to develop a different and more combative set of reflexes. On the basis of their entire history, it is logical to assume that they would encourage each other to do so by telling stories about people who had exhibited such reflexes in the past. In the final part of this book, though I will list a few important policies and principles for our coming acts of resistance, I want primarily to outline a set of reflexes that we might encourage each other to adopt.

15

Un-cancel the Future

In 2017 the luxury brand Calvin Klein launched a perfume called Obsessed, advertised by the British supermodel Kate Moss. When I first saw the advert, on the back of a glossy magazine, I was surprised at how young Kate Moss looked. On closer scrutiny I realized they had simply reused a photo of her from 1993.

This was logical because the new perfume was simply a 'reinvention' of a famous perfume launched twenty-five years before, called Obsession. 'It lived in our heads for so many years and became a touchstone of sensuality,' said the product's marketing boss. 'We thought about a scent that could reflect such an idea of memory and desire for today.'

Usually, if a company used a 25-year-old image to promote a product, we'd call it 'retro'. When marketers did that with pictures of Marilyn Monroe in the 1980s, the retro inference was clear. Likewise, when we watch an old *film noir*, or listen to a Billie Holiday track, or wear vintage fashion, we know we're consuming an aesthetic that has passed: we're indulging in controlled nostalgia.

But with the 2017 Obsessed advert, it was impossible to read it as retro. It looked modern, because in the quarter-century since the original advertisement appeared, so little in popular culture has changed. In the bars, cafés and hairdressers of the world, music produced in the 1980s is still played to create an ambience everybody can relate to. It is as if you played Glenn Miller to people in the 1970s, but they couldn't hear the difference between Swing and Punk.

The Italian philosopher Franco Berardi calls this phenomenon 'the slow cancellation of the future'. Once people subliminally bought the idea that neoliberalism was the final form of capitalism and that history had 'ended', popular culture entered a loop-cycle in which the idea of progress evaporated. Until around 1989 it had been normal to see new bands rejecting old styles and inventing new ones; cool kids suddenly appearing in clothes they'd improvised, making everyone else look old. Now everything became a montage of everything else. The cultural critic Mark Fisher summed up the feeling: 'Everyday life has sped up but culture has slowed down.'[1]

This absence of progress cascades over into politics. If we watch the original footage of police attacking Martin Luther King's march in Selma, Alabama, we understand that something good came out of the evil. Today's imagery of the alt-right marching with guns through Portland, Oregon, or the torture of prisoners in Syria, can feel instead like an endless, grotesque theatrical event, devoid of meaning.

The philosopher Fredric Jameson wrote that after the victory of neoliberalism, people found it easier to imagine the end of the world than the end of capitalism. But what if we could imagine the end of capitalism? Close your eyes for a moment and try it. Is it scary? What do you see?

Most likely you'll see the same utopia that has inspired Western thought since Aristotle: a community without poverty, in which property and hierarchy are unimportant, in which everyone has enough free time to develop their human potential and enough material resources to live, and in which the work is done by machines. The good life.

In the early twenty-first century, the means to liberate ourselves from work are close at hand. When you hear scare stories about robots or automated processes destroying half of all the jobs in the developed world, what it means is: we could be free of most physical work within a century. It means that the basic things we need to live – food, energy, transport, housing, healthcare and education – could become abundant enough to allow their provision outside the market, through direct collaboration with each other. Scarcity would exist increasingly in small pockets, dependent on expertise or natural resources.

In my book *Postcapitalism*, I argued that information technology has opened a new route beyond capitalism. Since the book came out, some of its proposals – such as the citizens' basic income, the provision of universal basic services or platform cooperatives to replace companies like Uber – have moved into the mainstream.

At the same time, thinkers like the Belarussian technology writer Evgeny Morozov have spelled out a clear and scary alternative outcome: digital feudalism. In this scenario technologically fuelled inequality takes off so hugely that, in place of the market, a new relationship develops between the tech companies and the mass of people that calls to mind the landowner–peasant power structure of the Middle Ages.

Wealth is extracted by the tech giants in alliance with the state, via owning and manipulating the data we produce. Most people can no longer meet their needs through work, because there is not enough of it; instead, they become bound to the technology providers in a form of servitude based on data.[2]

If we do end up with a kind of digital feudalism, the religion holding it together will not be medieval Christianity. It will be Kate Moss, forever young, advertising one new perfume after another, each a reinvention of the last. All the culture, fashion, music and art of the past 200 years will become 'samples': memories of a time when humanity cared about progress, liberation and the possibility of freedom, destined for reuse by people with little or no direct acquaintance with such ideals.

The capitals of digital feudalism would, of course, be Beijing, New Delhi and Moscow – because the immense power of the authoritarian regimes now existing would give these countries a head start when it comes to using artificial intelligence, surveillance and algorithmic control.

The rapid development of AI, together with Trump's offensive against the rules-based global order and China's emergence as a world power under Xi Jin Ping, makes digital feudalism a bigger danger than I originally thought. If it were to come into existence, one precondition for it would be that the robotics, AI and social media companies surrender their intellectual property to new, oligarchic states. In that sense, it would not really be a form of feudalism but a kind of 'second coming' for the bureaucratic collectivist nightmare that inspired Orwell's *Nineteen Eighty-Four*.

But the choice is still ours. If we want to take control of the technological possibilities before us, we need to describe the end-state we are aiming at and take action to clear away the obstacles to it. In order to un-cancel the future, we need to revive our reflexes for utopian thinking.

We need to understand first of all that capitalism is a complex, adaptive system which is losing its capacity to adapt. For more than 200 years, as technological progress made things cheaper and destroyed the need for certain skills, the system adapted by creating new needs and markets inside developed societies.

At the same time, capitalism has survived by using our planet as both a tap and a waste pipe: it has assumed the earth has unlimited capacities to provide raw materials and energy, and to absorb both waste and carbon. But the waste pipe is blocked and the tap is running dry. Climate change, pollution, resource depletion, demographic ageing and mass migration all look like serious 'external' shocks to the system, but are in fact long-term by-products of capitalism itself. And they have begun to feed off each other.

On top of that, information technology is limiting capitalism's ability to do the four things it has always relied on.

First, due to the specific nature of information technology, prices become difficult to form in a free market and profits become difficult to achieve.

Second, the existing technologies have potential to automate rapidly about half of all the job functions existing today, and – with further advances in robotics and AI – many more in the long term. Insofar as capitalism relies on the exploitation of workers, the prospect of a world in which work becomes increasingly optional disrupts its key dynamic.

Third, information technology creates network effects – new sources of utility, for example aggregated user data from a hospital or city transport system – which do not spontaneously appear as private property, and which are not owned in advance by either the capitalist, the worker or the consumer, but become the subject of a struggle.

Finally, digital technologies allow information to be democratized – removing the natural monopoly on distribution of knowledge that existed when it had to be spread via printed paper, or via the scarce radio waves of broadcasting systems and the rationed typewriters of the totalitarian state. This was a permanent feature of human life for 500 years and now it's gone.

In response to these four unique effects of infotech – on prices, automation, networks and availability – the market system has morphed rapidly, creating new organizations, laws and defence mechanisms. These include vast monopolies whose main aim is to suppress the free, competitive formation of prices, and to eradicate competition in entire swathes of the market. In turn they have pioneered strategies to extend artificial forms of ownership over information: extensions of copyright, complex legal obligations, intellectual property laws and 'non-compete agreements' that would have been pointless in the age of the analogue factory and the coal mine.

Far from rapidly automating production, advanced market economies are creating millions of jobs that do not need to exist – 'bullshit jobs' as the anthropologist David Graeber calls them. In the UK there are at the time of writing 20,000 hand car washes staffed mainly by migrant workers: twenty years ago they barely existed; and over the same period the market for automated car wash machines has collapsed.

Faced with network effects, tech monopolies have designed their business models to capture these positive spillovers in the form of economic rents. When you log into Facebook you are plugging into a machine for capturing the value produced by your everyday interactions. Ditto when you buy a smartphone contract. One by-product of this is that the most useful data in the world – from health to transport to the behavioural models that Facebook sold to Russian intelligence – are in private hands, not scrutinized by the state or by the public, and not open to social use.

Finally, to counteract the democratization of knowledge, corporations adopted the strategy of massive asymmetry, intellectual property capture and algorithmic control. As a Facebook user I am not even allowed to understand what Facebook knows about me, let alone what it knows about everyone else. Nor am I allowed to know what the algorithms are that guide content or advertisements towards me – nor what my aggregated data (in the form of 'synthetic populations') is being used for. It has, to borrow Fredric Jameson's phrase, become easier to imagine the end of the world than to imagine Mark Zuckerberg telling me exactly what his algorithms are designed to do.

As a result of the rise of information technology, we are now locked into a three-way struggle between the tech monopolies, the citizen and the state. This conflict is parallel to, and overlies, all the conflicts arising endemically from the economic failure of neoliberal capitalism.

As the tech monopolies grow in power they accelerate capitalism's transformation from a system based on production to one based on rent-seeking. Though some innovation is pursued with the goal of automating jobs that don't need to exist, much of commercial innovation is aimed simply at creating new monopoly opportunities; new mechanisms to subvert the democratization of knowledge; new asymmetries of information; new ways of destroying old social structures, like the business ecosystem of the taxicab industry, or commercial office space.

No serious faction within the Western corporate elite wants to challenge the rent-seeking models that are strangling the economy. The elite of the emerging markets is dependent on them – and so, increasingly, are a caste of global lawyers, bankers, politicians and service industries for the super-rich. This in itself is a symptom of the dead end the capitalist mode of production is approaching.

However, parts of the libertarian right do have a solution: to abolish the state and all welfare protections, turning the digital economy into a gigantic, atomized market, in which there are no central banks, only Bitcoin; no states, only blockchain contracts and therefore no human rights. The ultra-right's utopian reflexes are strong and they lead to a society of total algorithmic control.

The combination of monopoly, precarious work, artificial scarcity and information secrecy with long-term economic hangovers from the failure of neoliberalism makes it likely that the current model of capitalism will suffer repeated breakdowns. However, human freedom is closer than at any time in history because thinking machines are unique. They create new utility for free, and on a vast scale.

The solution is that we consciously design a new global social system to utilize the capabilities of automation, reduce the amount of work needed to keep us alive on the planet and in the process stabilize the planet's ecosystem. Critical to that project will be the regulation of artificial intelligence, the protection of data rights and resisting the control of human beings by algorithms.

*

The end-state we should try to achieve is technological abundance: a world in which machines do most of the work, even most of the innovation; in which our massively expanded leisure time allows us to experience a rich cultural life; and in which our economic activity moves into harmony with what the earth can sustain.

To get there, I propose four strategic projects, each matched to one of the effects information technology has created within capitalism.

1. To combat monopolies and price-fixing: break up the information monopolies and promote the socialization of basic digital infrastructure, in the form of non-profit companies or state-owned utilities similar to the energy grid.
2. To combat precarious work and stagnant wages: accelerate automation by de-linking work from wages. This involves paying everyone a citizens' basic income, out of taxation, plus the universal provision of four basic services – healthcare, transport, education and housing – either ultra-cheap or free. These measures should act as a transitional subsidy to offset the impact of rapidly automating the world.
3. To combat rent-seeking: legislate to make data into a public good, while giving ultimate control of how each person's data is used to the individual, not the state. Suppress all business models based on rent-seeking; indeed make the seeking of economic rent socially unacceptable.
4. To fight information hoarding: outlaw all business models based on asymmetric access to information. I should have the right to know, and to see, what any state, any bank or any social media company knows about me. I should have the right to delete the information, to correct it and to limit its use. I should have the right to know if an algorithm is being used to control, monitor or predict my behaviour. I should have the right to know if an artificial intelligence is being used on the other side of a transaction, game or conversation.

These four strategies are designed to unlock the economic power of the new information technologies that are being developed now, and whose expected impacts on medicine, robotics and urban living are often labelled the Fourth Industrial Revolution. They are explicitly designed to stop info-capitalism morphing into digital feudalism, and to prevent the emergence of digital anarchy; and to lay the basis for a transition to a diverse collaborative, non-market, collectively owned economic model.

The transition will be slow; the forces of postcapitalism will have to build up inside what's left of capitalism. Since no system disappears before producing all the technologies, techniques and social forms latent within it, we should not attempt any kind of 'forced march' towards abundance and cooperation, but instead nurture its earliest forms carefully, in the spaces available within the present: non-profits, collaborative production, the peer-to-peer economy and open source software and standards.

On that principle, one of the most important aims will be to leave room for real entrepreneurship (as opposed to the tax-dodging, rent-seeking and slave-driving activity that passes for it under neo-liberalism); for market-led innovation, for innovative partnerships between the public and private sectors in which the state consciously cedes opportunities to create value to the private sector.

By the same principle, planning will have to take a very different form than under twentieth-century Stalinism and state capitalism. It has to move beyond the urban and infrastructure design and 'industrial policy' strategies used in market-dominated societies. The most important tool of the planner should be a complex digital model of the economy, society and ecosphere, at a local, national and global level.

The model needs to do what the plan could never do: anticipate the complexities, the feedback loops, the social and environmental impacts of government decisions or industrial strategies over decades – and present the results simply enough for voters to make an informed choice. Models should be tools for the electorate, not just technocrats, to experiment with: to explore and imagine possibilities and test them, thereby massively increasing democratic participation and access to social knowledge.

From basic income to the circular economy, to the creation of platform co-ops or information as a public good, most of these ideas are already in circulation. Some are being enthusiastically adopted by city governments or in niche parts of big corporations. But as its opponents have acknowledged, the most likely vehicles for a postcapitalist solution to the present crisis have emerged in the form of radical left and green parties, or radical left tendencies inside liberal and social democratic parties.

The first concrete action that you can take if you agree with the analysis in this book is to start embedding this four-part approach as a reflex – within parties, trade unions, communities and social organizations: to make stuff cheap or free, to de-link work from wages, to promote data as a public good and suppress the right of corporations to monopolize information.

But having outlined the goal, we need to take concrete steps towards it. In Britain, for example, after ten years of austerity and thirty years of free-market devastation, putting things right would require actions borrowed from the old programme of the left.

We might need to take public services that are failing and ripping off their customers into public ownership; to raise wages by outlawing precarious work and empowering trade unions; to tax, borrow and spend in order to create new, vibrant human-centred public services, public space and a modern infrastructure.

But we cannot allow this to become confused with a return to the state-capitalist project or, even worse, the Stalinist socialism allied to it. When we build thousands of new social housing units, we should think about community control; carbon-neutral building methods and materials; permanent rent controls; and the creation of sustainable, mixed communities with enough public space to support a flourishing democratic culture.

By the same token, if we believe that only a community with a strong institutional and ethical life can resist control by the new technologies we have invented, every action – by a government or a progressive political activist – has to build this capacity, or at least not diminish it.

But the art of radical politics today lies not only in crafting a vision of the future and a transition path towards it. It also involves a political struggle against the authoritarian right – and as the rise of the alt-right suggests, this too demands a different set of reflexes from the ones we've been using.

16

React to the Danger

'Don't you feel,' the poet Stephen Spender asked George Orwell, 'that, any time during the past ten years, you have been able to foretell events better than, say, the Cabinet?'

It was June 1940. London's railway stations were awash with troops evacuated from Dunkirk. The foreign policy of the British elite was in ruins, half the Cabinet wavering towards a deal with Hitler, and the UK was substantially defenceless.

Orwell, who had been predicting war with Germany since 1936, replied: 'Where I feel that people like us understand the situation better than so-called experts is not in any power to foretell specific events, but in the power to grasp *what kind of world* we are living in.'[1]

It is a power that in our time has evaded the liberal centre. As the free-market economy fails, as figures like Trump run rings around their mainstream rivals, the intellectuals and politicians of the establishment look as clueless as they did in Orwell's era. They thought they were living in a system that stabilized itself. Instead, it destabilized itself.

But Orwell's words advocate a powerful human reflex: to understand *what kind of world* you are living in. Applied to today, it means we have to accept that the build-up of tension within the system is going to lead to a blow-up, even if we can't predict in what form. The elites who've run things for the past four decades are spineless, capable of any form of compromise with the authoritarian right, and very easily separated from their democratic principles. This is powerful knowledge in itself, so long as it doesn't paralyse you.

The dynamics that drove Brexit, put Trump in power, and allowed overtly racist parties to top the polls in Italy, Sweden, Hungary and the Netherlands will not quietly subside. The clamour of the alt-right for American Civil War 2.0 will not evaporate. The imagery of tortured prisoners and shattered cities will not grow stale.

Hans Mommsen, a left-wing German historian who studied Hitler's rise to power, described the interaction between the Nazi party, the German business class and the civil service as 'cumulative radicalization'. The results are known. Today, though the dynamics of our crisis differ from those in the 1930s, 'knowing what kind of world we live in' means understanding how elites, fascists and bureaucrats are undergoing a cumulative radicalization towards racism, xenophobia and the curtailment of democratic rights.

That in turn means we have to formulate a strategy to stop this radicalization, even if it requires putting parts of our own project on hold. Put bluntly, it's a question of urgent versus important: for many progressive people, what's been important to them in the past is not the same as what is urgent now.

Though the barbarism of the 1930s was worse than ours, it was hidden from most people by the monopoly the elites had over information. Even during the Second World War there was almost complete popular detachment from, and ignorance of, the big events until they hit people. At the height of the Dunkirk crisis, Orwell's diary records:

> People talk a little more of the war, but very little. As always hitherto, it is impossible to overhear any comments on it in pubs, etc. Last night, Eileen and I went to the pub to hear the 9 o'clock news. The barmaid was not going to have turned it on if we had not asked her, and to all appearances nobody listened.[2]

That was the same day Churchill had made his speech to the Cabinet, pledging to fight until 'each one of us lies choking in his own blood upon the ground',[3] but the drinkers in Orwell's pub knew almost nothing about the events shaping their lives.

We, by contrast, receive every petulant thought of Donald Trump via Twitter. We follow air strikes and terror attacks in real time. We have seen, and cannot unsee, what torturers do, and what a beheading looks like. As a result, our levels of fear and anxiety are probably higher than in the run-up to the Second World War, our fight and flight reflexes highly attuned.

Writers from the generation that survived fascism – Arendt, Fromm and Orwell himself – were fascinated by the mass psychology of fascism. Today, preventing fascist psychology from becoming mass is one of the most important tasks for progressives and democrats. In many advanced countries we are faced with something new – a fascist project that is simultaneously better informed and more knowingly implicated in what it wants to achieve, and armed with the ability to speak in subtexts, jokes, memes and conceits like Kekistan.

In the 1930s, Mommsen pointed out, it was the German elite's attacks on constitutional government that produced the conditions in which Hitler flourished. Today, across the G20, curtailments on democracy and the rule of law proliferate: Trump's executive orders banning Muslims from entering America or pardoning crooks; the Spanish state's violent suppression of the Catalan independence struggle; Poland's suspension of constitutional protections for its judiciary; the rigging and manipulation of the British Brexit referendum using Russian money; the widespread trope among authoritarian nationalist politicians and newspapers of attacking the judiciary as 'enemies of the people'.

The clearest danger today is not that fascist movements become big enough to win elections or seize power; it is that they create a shared mindspace with mainstream conservatives which erodes the willingness of the centre right to resist their demands, and which instead becomes an excuse for eroding constitutional democracy.

A case study in how this happens was Roy Moore's run for the US senate in Alabama in 2017, on the Republican ticket. Moore had been removed twice from the position of the state's chief justice for refusing to uphold constitutional law on the separation of Church and state, and the legalization of same-sex marriage. After his campaign for the senate began, nine women came forward to allege he had made sexual advances to them, two while they were below the age of consent. Moore has denied the allegations.

Trump publicly supported Moore. Steve Bannon spoke at his rallies, but his core activists came from two networks: the League of the South, a white nationalist group that wants to re-form the

Confederacy; and extreme anti-abortion activists who justify killing doctors who work at abortion clinics.[4] Despite the massive reputational hit to the party, the Republican National Committee endorsed Moore. His narrow defeat could not obliterate the seriousness of the problem: the most influential conservative party in the world had been sucked into the game of race war and violent misogyny.

Moore's election run in Alabama showed not that a white supremacist, violent misogynist and anti-constitutional movement is about to triumph in the USA, but that this option is on the agenda, that it informs people's fantasies, that mainstream conservatism's defences against it are weak, and that the shared mindspace between conservatives and fascists is currently pulling politics towards the extreme.

The parallels with the 1930s hold important lessons for how we fight this. The first is that as far as possible the radical left and the liberal centre should stop fighting each other.

In France, where on 6 February 1934 hundreds of thousands of people took part in a far-right demo in Paris, trying to overthrow the government, working-class activists forced the communists and the more moderate socialists into practical unity by simply building it at grassroots level.[5] In Spain two years later, when the fascist General Franco tried to seize power from an elected liberal government, unity in action stopped him. Only a three-year civil war, with military intervention by both Nazi Germany and fascist Italy, allowed Franco to take control.

Today, faced with this new threat from the right, the liberal centre is making exactly the same demands on the radical left, on greens and trade unions as it made of antifascist workers in the 1930s: forget the fight for social justice and form up behind our leadership, with our values, on our terms and in order to defend our failed neoliberal project.

The result so far has been mostly negative. It worked briefly for Emmanuel Macron but failed for Hillary Clinton, failed for the Remain campaign in Britain and failed to stop the Austrian conservative party forming a coalition with the far right. In many countries with weak democratic cultures – in Eastern Europe for example – it is a non-starter anyway. Meanwhile, many of those on the left just cannot get their heads around the idea that the main enemy has changed.

To regroup, we have to develop the reflexes of both the left and centre towards finding common ground. This has to start by acknowledging that the differences are maybe bigger than in the 1930s. Back then, the project of the left was simply a harder version of the project of the liberal centre: state ownership and control of industry, welfare programmes and trade protectionism.

Today, individuals from the educated, progressive and secular half of the population are often grouped around a last-ditch defence of globalism and deregulation, while a more radical left is gathered around a programme of social justice, de-carbonization and the overthrow of neoliberalism. At its worst, the liberal centre has more or less given up trying to address the economic concerns of the workers who vote for right-wing populist parties, and can easily lapse into treating them with contempt.

But what the neoliberal centre and the radical left share is the need to defend democracy and the rule of law. In any given country, I would place that at the centre of any attempt to form tactical alliances to defeat the right.

Second, we need to develop the strategies that prevent the convergence of conservatism, fascism and the state bureaucracy into a common, authoritarian project.

This means, wherever possible, isolating and suppressing fascism. Though fascist groups remain tiny in most countries they represent the public promise of genocide. Groups who march through American cities carrying assault rifles and chanting 'Jews will not replace us' are not doing this as a gesture: they are doing it in preparation for murderous attacks on minorities and to create the chaos into which their elite allies will step, armed with emergency powers, to suspend constitutional democracy.

To fight the possibility with intelligence needs more than the traditional tactics of 'anti-fa' confronting the small fascist groups on the streets. It requires the progressive half of society to force the executive and the judiciary to defend the rule of law and to maintain the state's own monopoly on armed force.

Unfortunately, of all the advanced democracies, the one most systemically weak when it comes to this is the biggest: the USA. Its judiciary, in the neoliberal era, has become highly politicized – not just via rival political appointments to the Supreme Court, but by the politicized use of federal prosecutions. The state's monopoly of armed force, already undermined by the abuse of the Second Amendment, is rapidly eroding as the alt-right militias, and the right-wing 'constitutional sheriffs' who tolerate them, create alternative armed bodies of people.

In his account of the rise of Nazism, Mommsen pointed to the emergence of legally tolerated militias numbering up to one million people, financed by landlords and industrialists, in creating a prevailing atmosphere of disorder and informal violence. Hitler's brownshirts were only the most unruly element in a much wider ecosystem of armed groups, some of which inter-operated with the police.

Except in countries like Germany with tough restrictions on neo-Nazi groups, one of the big unacknowledged weaknesses of Western democracies today is the preparedness of their law enforcement systems to tolerate localized fascist violence, widespread and coordinated hate speech, and far-right infiltration of the police and armed forces. Reversing that situation through legislation and executive action is urgent. But the unwillingness of liberal centrist governments to do so is yet another signifier of *what kind of world* we are living in.

When the British Home Secretary Sajid Javid called left-wing members of the Labour Party 'neofascist', in the same month that thousands of actual fascists rampaged through central London pelting his own police officers with missiles, it is logical to conclude that – when the crunch comes – large parts of centrist conservatism will mount zero defensive effort against the authoritarian right wing.

However, where at all possible, as far as concerns the newly invigorated right-wing populist parties – such as the Freedom Party in the Netherlands, Italy's Lega and the AfD in Germany – the most effective progressive tactic is to keep them organizationally isolated: trapped between official conservatism and the alt-fascist or neofascist right. It was the Republican Party's failure to do this which allowed it to pass into the hands of the Tea Party.

Finally, though all the sociological evidence says the authoritarian right is being boosted primarily by cultural insecurity, not economic stagnation, the historical record shows that economics can still help us deflate right-wing populism.

Anybody who has tried to debate with xenophobes and ethnic nationalists on the doorstep knows they have good days and bad days. On a good day they are mainly angry about the lack of jobs, or reports that allege that migrants drive down wages. On a bad day they call people from ethnic minorities 'cockroaches' and will claim that they want an end to migration 'even if the economy collapses'. Likewise, on a good day nationalists such as the Law and Justice Party, which governs Poland at time of writing, are embarrassed that around 8 per cent of their vote comes from outright antisemitic fascists; on a bad day they are glad of it.

The answer is an assertive programme of economic expansion, with the positive effects front-loaded into the kind of communities where authoritarian nationalism is flourishing. If we are to make the right-wing offensive go away, the left and the liberal centre need to make a demonstrative break with the failed economic model of neoliberalism. As Jeremy Corbyn's radical Labour manifesto showed in the snap British election of June 2017, even a rhetorical break can be good enough to make supporters of right-wing populist parties like UKIP switch straight back to voting left.

In gritty working-class towns I, and those campaigning around me, personally experienced white male workers saying: 'we're coming back; all it needed was for somebody to show they cared about us'. That doesn't absolve, or solve, their flakiness towards right-wing politics. But every far-right lawmaker dissuaded from standing, every local group disbanded, every racist who gets off the street and goes back to writing green-ink letters to their local paper, is an achievement.

The left and the liberal centre cannot give the right-wing populist voters what they want most: a return to social conservatism, the revival of white privilege and draconian immigration policies. But for precisely this reason we need to double down on the things we can offer: jobs, investment, training, infrastructure and a narrative of hope.

Though there is no agreed name for it, and no single institution controls it, after 2014 a clear progressive alternative to neoliberalism began to emerge. In Greece, Syriza launched a six-month resistance to European austerity before surrendering in July 2015; the Podemos Unidos party in Spain, and its city-level alliances such as En Comu in

Barcelona, began to score consistently 20 per cent in the polls. The coalition government in Portugal, including socialists and the far left; the Bernie Sanders faction inside the US Democratic Party; and the movement around Jeremy Corbyn which took control of the UK Labour Party, all testify to the crystallization of a new left politics aimed at gaining political power.

In almost all cases the critical factor pushing these left parties and movements out of their ghetto of political purism and towards power was the switch by tens of thousands of networked activists from the 2011 period into party politics. They brought with them vision, energy, organizational skill – and an ability to connect old parties with a millennial generation that had in many countries switched off from politics.

If you look at the places where such a movement could have emerged but did not – Ireland, France or Iceland for example – the common factors are (a) the absolute deadness of traditional social democratic parties and (b) locally important divisions on the left which meant its constituent parts could not unite around a clear, single project.

In the Corbyn movement a further important factor was that hundreds of single-issue activists, with huge amounts of social capital sunk into lifelong obsessions – such as Palestine, climate change or even academic critical theory – were prepared to deprioritize these issues and collaborate to do one thing, over an extended period: transform the Labour Party into a tool for ending neoliberalism.

In each of the new left parties, out of the practice of taking small steps to resist while defending and enriching democracy, has come a rough-and-ready ideology. It is so pervasive, despite its constituent parts being so diverse, that it has to be understood as an essential feature of the world we live in. Let's express it as a formula: networked activism, plus a focus on party politics to achieve state power, plus relentless focus on the issues, language and concerns of ordinary people are the basic ingredients of the new left project.

In response to Orwell's injunction – that politically active people need to understand 'what kind of world we are living in' – the answer is: a world in which either the left reinvents itself as a popular social movement for radical change, or democracy dies.

17

Refuse Machine Control

Picture this: in a dusty town on the Mediterranean coast, a man walks up to the local HQ of law enforcement, clutching the holy book of a fanatical religion and ranting about its forbidden message. The authorities tell him: obey the law, worship the official religion, hand over your banned books – and we'll let you follow your crazy beliefs in secret. But he does not want secrecy, he wants martyrdom. He is jailed, tortured and eventually executed.

This is not the story of a twenty-first-century jihadi but of Euplus of Catania, a Christian martyr executed in 304 CE. Euplus was one of up to 3,500 people killed when the Emperor Diocletian, in an effort to suppress Christianity, forced everybody in the Roman Empire to make a public sacrifice to the pagan gods. His aim was not to impose beliefs but behaviours. Conform to the social norms, said the authorities, and you can believe what you like.

But the early Christians did not conform: they chose defiance. And although some were killed, most were not. The repression fizzled out. Christianity became legal ten years later and by the year 380 it had – in a remarkable turnaround – become the official religion of the Roman Empire.

There are only two explanations for Christianity's rapid spiritual hegemony: the miraculous one promulgated by the Church, which gives the martyrs magical powers emanating from God; and a materialist one that tries to situate this huge event amid the struggles over property, power and land as the Roman Empire declined. Marxism has always struggled to give such an account, because its historians were always looking for Christianity's narrow roots in economic struggles rather than its material force as an expression of human values.

Just why Christianity swept Europe and the Near East as the Roman Empire failed becomes clearer once we understand the historical period it concluded. Between 800 BCE and 200 BCE most of the great human-centric religions emerged – an era the philosopher Karl Jaspers called the 'Axial Age'. Confucius, Buddha, Lao-tse, Zoroaster and many prophets of Judaism all lived around that time. Most of them performed the same basic social function: they were wandering ascetic thinkers who tried to influence the powerful rulers of warring city states and inculcate values of restraint based on claims about human nature. What happened in these few hundred years, wrote Jaspers, 'was that man became aware of existence as a whole, of his self, and of his limitations'.[1]

But this Axial Age doesn't quite fit into strict materialist accounts of history. It comes after the great civilizations of the Bronze Age and before the great trading empires of the Iron Age. For Marx, history was categorized through 'modes of production': classical antiquity, Asiatic despotism, feudalism and capitalism. Because the Axial Age idea spans both classical Greece and the Zhou dynasty in China, many left-wing historians have filed Jaspers's insight as 'interesting but unimportant'.

However, the anthropologist David Graeber has offered a plausible materialist explanation for the Axial Age: it almost exactly coincides with the rise of coinage. Though money had existed for thousands of years, small pieces of metal stamped with the crest of a king or city state appeared almost exactly at the same time as Confucianism, Buddhism and human-centric Greek philosophy, and in the same places: on the Yellow River, the Ganges and in the eastern Mediterranean.

Graeber says what emerged after 800 BCE was a 'military-coinage-slave complex', which forms the common basis for very different (and largely unlinked) city states across China, India and ancient Greece. Coins were needed to pay highly trained standing armies; they facilitated the creation of market-oriented societies and states whose dynamism was linked to wars of conquest and the possession of slaves. And this led to a new concept of the world, in which the material wealth of communities was understood as the highest good. If Graeber is right, the system that is born with coinage should be seen as a distinct mode of production in itself, though shorter lived than the thousands of years of slavery Marx lumped together in one system.

'Everywhere we see the military-coinage-slavery complex emerge, we also see the birth of materialist philosophies,' writes Graeber.[2] You get mass literacy, you get new humanistic concepts such as Aristotle's 'good life' or the Confucian ideal of *ren*, best translated as 'cultivated humane-ness' though often used interchangeably with 'virtue'. And eventually you get mass popular movements that use rationalism to challenge the inherited power of rulers.

If so, the role of Christianity in collapsing the Roman Empire throws light on our own time. Orthodox Marxist historians thought the slave system, which was at the core of the Roman economy, collapsed because it could not improve the productivity of the land. The elite detached themselves from an economic model based on the prosperity of citizens, instead building a state that consumed more and more of the surplus and employed more and more slaves. Eventually, landowners switched to employing bonded farm labourers, called *coloni*, and the market for slaves collapsed because the bonded labourers proved more efficient. At this point of maximum weakness, the Germanic tribes swept in and, amid the wreckage, their own system of bonded labour fused with that of the Roman farmers to create early feudalism.

That account is economically plausible. But if we understand the Axial Age to have been based on the implicit promise of a human-centric economy, we can understand why its point of failure triggered the rise of an ideology that insisted slaves and free-men were equal. Christianity was a call for a human-centred society within a system that had promised it but could no longer deliver it. It was a call for a morality more powerful than the laws of a state which had become increasingly barbaric towards its citizens, who themselves were increasingly a minority of the population; an elite who were extracting a surplus in ways that seemed archaic, inefficient and inhuman.

By limiting the story of the end of Rome to a narrative of economic collapse, the orthodox Marxists who studied it in the twentieth century blinded themselves to what can happen when an idea becomes a material force. As an ideology of revolt and of small-scale humanist, law-based communities, Christianity helped create the economy that would replace the slave mode of production. With its emphasis on equality of individuals before religious law – a law higher than that handed down by any local ruler, Christianity helped form a new economic model to replace the slave system, even though in practice it sometimes tolerated slavery.

Today we know that what replaced Rome was not simply a 'Dark Age' of chaos and warfare but a thriving, decentralized economy based on bonded labour, which produced its own rich culture, one focused on artefacts and manuscripts rather than large buildings, and its own attempts to revive classical learning.

Between 300 and 500 CE a mental, moral, ethical and behavioural revolt centred on a humanist religion helped kill off the parasitic economy of ancient Rome. Though I do not advocate martyrdom or a return to Christian theology, that is an interesting lesson for us. The Christian 'revolution' of the fourth century happened because large numbers of people refused to go on obeying empty rituals.

The events of the fourth century CE show that, when a system is heavily dependent on people following control routines ordained by the elite, refusing to conform to those routines can have revolutionary consequences. It follows from this that this kind of refusal could be very powerful in our fight to replace neoliberalism, prevent fascism and resist algorithmic control. The basic reflex we need to cultivate is the power to refuse.

If you want to see how easily humans can be controlled by an algorithm, think of an airport. Normal behaviour stops as you enter the building: after that, strict rules apply. The check-in establishes your identity, and the security scan is a mini-algorithm of its own: people move as ordered, yank their laptops from their bags as if life depended on it; go into the sub-routine of a body search. As your passport is scanned, all relevant facts the state knows about you are checked, while a face-recognition sub-routine makes sure you are the same person who checked in a few minutes before. At the gate, the algorithms of economic privilege begin: the rich board first, the poor board last.

The natural human reaction to this level of directed behaviour is tension. But regular travellers learn that anxiety is pointless. As a journalist covering global stories I learned to expect arbitrary disruptions, heavy and impersonal control. I learned not to care if my luggage got lost, if flights were cancelled, or if irate airport workers yelled at me. Like millions of others subject to algorithmic control, I learned to go with the flow.

The problem is that information technology is turning more and more of our everyday lives into the equivalent of an airport. As with airports, the main drivers are the needs of corporations and states. Unlike in airports, most of the algorithmic control being exerted on our brains and bodies is not obvious: it is not publicly understood, nor is it being regulated, nor are its ethical dimensions being properly considered by society.

Let's start by understanding what an algorithm is. It's the use of logic to turn complicated situations into a series of yes/no questions, and to issue instructions according to the answers. The algorithms at work in the airport are all in their different ways asking the same question: 'is this person safe to allow on a plane?' An algorithm is logic plus control.

To understand how quickly algorithms have evaded scrutiny and accountability, log onto YouTube and search for 'finger family'. You will be presented with up to 17 million 'different' videos aimed at nursery-age children, which all look much the same. Another You-Tube channel devoted to hours-long videos of a person unwrapping chocolate eggs or unboxing toys has 3.7 million subscribers and, by 2017, 6 billion views.[3] The titles of the videos themselves are literally meaningless: a jumble of keywords, which are not designed to attract human attention but the attention of the algorithm that serves up the next video in the queue.

The videos are chosen via algorithm, promoted via algorithm and even created automatically using software that copies and pastes the same mesmerizing rubbish from one file to the next.

Once you put a child in front of such a YouTube channel you are surrendering control of their viewing to an algorithm. Another machine – a bot – is crawling across YouTube pretending to click on certain videos, in order to drive them up the rankings. Meanwhile, another bot is leaving computer-generated comments, again to boost the ratings.

The machine is choosing what your child can watch, and quite possibly permanently shaping how they perceive the world. James Bridle, a British artist who studied the effects of digital intelligence, called this 'infrastructural violence' – a form of coercion so invisible that we don't have words to talk about it – and which only Google and its subsidiary YouTube have the knowledge needed to protect us from.[4]

We are beginning to understand the most obvious dangers of algorithms. They spontaneously reproduce existing human bias – such as software used in America to assess job applicants, which was producing racial discrimination. When used to assess teacher performance in the USA, algorithms produced results that were so badly skewed that teachers concluded they were being used simply to instil fear and discipline into the workforce. Many states and cities abandoned the software.[5]

In these early case studies, the common response to a bad algorithm is to think we're dealing only with a technological glitch that can be corrected. But the anti-human use of algorithms is almost always driven by economics, not technological opportunism or neglect.

The proliferation of algorithmic controls is a response to the four big effects of information technology I outlined in Chapter 15. The collapse of computer production costs of information goods makes it cheap to automate the production of animated videos. The creation of giant monopolies like YouTube allows the authors of kid videos to make money out of tiny slivers of advertising revenue. The massive reliance on information asymmetry blinds us to what's being done to our children's brains.

To mandate the ethical use of algorithms, with obligatory disclosure and opt-outs, would not only challenge the tech companies' business models, it would challenge the current dynamics of capitalism, and call into question the multibillion-dollar market valuations of key companies.

Such resistance has begun. Teachers' strikes, class-action lawsuits and the creation of ethical standards by bodies like Britain's IEEE (Institute of Electrical and Electronics Engineers) are twenty-first-century equivalents of what happened in the first fifty years of the factory economy, when class struggle and government regulation forced factory owners to stop polluting rivers and killing children through overwork. The question is: 'where will this resistance lead?'

As we resist algorithmic controls we need to do so in the name of more than safety or correcting bias: we do so in the name of our essential human characteristic, the species being.

Everything that reduces our conscious control over our work environment, or our rational choice or our freedom has to be resisted. Not in the name of technophobia, but in pursuit of better machinery; better, more transparent algorithms; more control. But in resisting the algorithm, we also need to understand the economic control mechanisms capitalism has become reliant on.

One of the most depressing rituals to emerge during the past twenty years is the TED Talk. The person giving the talk is unknown to most people but their delivery is slickly believable, and they're often filmed from below to make them look authoritative. The subject of the talks is also depressingly constant: how to take advantage of human weakness.

The top twenty-five TED Talks of all time include 'How great leaders inspire action'; 'How to speak so people want to listen'; a guide to the 'invisible forces that motivate everyone's actions'; 'How to spot a liar' and a professional pickpocket demonstrating how to 'swipe a wallet and leave it on its owner's shoulder while they remain clueless'.[6] Billed as a collection of 'ideas worth spreading', the TED Talk is actually devoted to spreading a single idea: that human judgement is fallible and behavioural economics can allow a few smart people to make a lot of money.

While they were busy destroying the welfare state and replacing human-centred policies with inflation targets, the neoliberals ignored behavioural economics. Their credo was: only markets are rational. As a result, those who had tried to study the actual behaviour of individuals in market situations, usually coming from the discipline of psychology, were sidelined.

The take-off point for behavioural economics is usually identified as an academic paper published in 1980 by US economist Richard Thaler called 'Towards a positive theory of consumer choice', based on the work of future Nobel Prize-winner Daniel Kahneman. But the practical take-off for the new discipline happened much later: after the dotcom crash, when businesses and governments began to notice people were refusing to conform to market imperatives as neoliberal theory said they should.

Offered crappy warranty deals by monopolistic sellers of refrigerators and microwaves, we bought them when we should have rejected them on value terms. Offered endowment mortgages that would never pay off the original debt, again we bought them. Offered guaranteed incomes by pension funds trading in highly volatile stock markets, we asked, gullibly: 'where do I sign?' At the height of the neoliberal era, people were engaged in irrational market choices on a vast scale.

In the ideal market imagined by neoliberal purists, these rip-off warranties, mortgages and too risky pension plans should be eradicated by competition. But capitalism had become an opaque system of monopolies aligned to state power, in which the efforts of a lobbyist in Brussels or Washington can more than offset the power of competition.

Thaler's 2008 book *Nudge*, co-authored with lawyer Cass Sunstein, gave birth to the strategy of 'libertarian paternalism', which was adopted by many governments as the free-market logic began to fail. *Nudge* became the inspiration for numerous policy units, including in Obama's White House and 10 Downing Street, whose aim was to shape the behaviour of citizens to help them conform to the market conditions in which they live.

In 2013, for example, noticing that poor students tended not to apply to Britain's top universities (known collectively as the Russell Group), the UK government Behavioural Insights Team launched a *Nudge*-inspired experiment. They sent 11,000 school students a letter from a student at a top university, pointing out that it can be cheaper to attend such colleges because they offer better financial support for poor applicants.

In 2017 the media showcased the positive results. Those who received letters were more likely to apply, to be offered a place and to accept it – although, as the researchers pointed out, not likely enough to be statistically significant. For the overall cost of £45 per student, 222 people had apparently been saved from attending a low-ranked university and had gained a place at a more prestigious one.[7] The *Economist* magazine gushed that 'the approach was less heavy-handed

than imposing quotas for poorer pupils, an option previous governments had considered'.[8]

Read the actual research however, and the authors state: 'It is note-
worthy that although our interventions have been successful on one
margin, we see no overall effects on the likelihood of students applying to university.'

And that is because in 2010 the Conservative–Liberal coalition
government had hiked student fees from £3,000 to £9,000 per year,
and had then privatized the student loans company, which was now
allowed to charge students interest rates of 6 per cent against a central bank base rate of 0.25 per cent.[9]

Neoliberals love the 'nudge' strategy because – as its authors
intended – it is a substitute for higher taxes, heavier regulations and
laws. In addition, the libertarian aspect of it places ultimate responsibility for the outcome on the individual – in this case, the
poverty-stricken sixteen-year-old making pizza for their siblings in a
cramped council flat, while trying to imagine how they might cope
with life at an elite university.

A much simpler strategy, and a clearer signal to society, would be
to force elite universities to adopt quotas for kids from poor back-
grounds (as some have done voluntarily). Even better, as the Labour
Party promised in the 2017 UK general election, would be to make
university tuition free. The most socially just strategy of all would
attack poverty at its roots by providing a generous welfare state.

Since none of this is conceivable to the neoliberal elite, we are left
with nudges. On some days it can feel as if your entire life is being
nudged. It happens at the coffee shop, where coerced and underpaid
workers are forced to smile at you and suggest a bigger doughnut; at
work, where the entire management team gets sent on motivational
courses and keeps saying 'let's think positively' even as the company
goes down the tubes.

Nudge strategies exist because markets are unfair – yet they do not
cure the unfairness of markets. They require us instead to participate
in the illusion that markets work. They require us to collude in the
presumption that heavy regulation is bad and 'choice architecture' is
good. They force us, in short, to reproduce neoliberalism consciously,
as if it 'ought to' work, because the theory that it works spontaneously

has failed. In this way, behavioural economics has become the twenty-first-century equivalent of the Roman demand to 'sacrifice to the gods'. Believe what you like in private, governments and corporations tell us, but in the public, commercial world please behave according to our religion.

So one of the most powerful things we can do is reject the old behavioural control systems at an individual level. Some already do: they buy only FairTrade coffee; they wear expensive jeans hand made in Wales rather than cheap ones mass produced in Bangladesh; they get out of taxicabs driven by racists and slam the door. Up to now neoliberalism is not worried by this – its response has been to co-opt ethical consumption, much as the Roman Empire tried to co-opt Christianity in the decades after Diocletian's massacre.

So to create a behavioural tipping point we have to raise our rejection of imposed market values to a new level. You can see how it might escalate. Customers refusing to use automated checkout machines and forcing supermarkets to employ humans; people using their consumer rights aggressively; above all loudly refusing to be nudged.

But if this form of resistance takes off, two things will happen.

First, we will start to find each other. If you stand in a long queue, waiting for the only human checkout employee and rejecting the automated checkout, sometimes you will meet a fellow customer who is irate. So you remind them that it's not the server's fault that they are waiting in line, but the fault of the multibillion dollar supermarket chain. As you do, a third person may say: 'that's just what I was thinking'. And now you have something more than isolated discontent.

Second, as we continue to make tiny acts of rebellion, we will think: this could be solved much more easily by a combination of better technology and by passing different laws. Our micro-scale resistance will lead us towards a society-wide project.

In drawing this parallel with early Christianity, I know I am inviting people to characterize the project of radical humanism as quasi-religious. I do not intend it to be. But the fact remains that the specific character of exploitation in the neoliberal era, which sucks value from us in the workplace, the pub, the home, the coffee bar, the sports field and the bedroom, is invasive and can be resisted by refusing to perform as the system demands.

As late as the 1970s, my generation could believe that resisting capitalism was something you did by pulling on a big lever to make a big boulder – the labour movement – move. Today, the more efficient way to start the avalanche is to be a little stone and start rolling.

18

Reject the Thoughts of Xi Jin Ping

What I remember most about the inauguration of Xi Jin Ping in November 2012 was the empty seats. In Beijing's Great Hall of the People, the entire ground floor had been reserved for 3,000 delegates to the 18th Congress of the Communist Party of China. We, the journalists, had free run of the first balcony, deserted except for ourselves and a large military band. The second tier, a balcony with 2,500 seats, was entirely empty. This was a great hall all right, but without the people.

Jiang Zemin, the man who oversaw the marketization of China and the creation of a cheap labour force in its export factories, sat grumpily on the platform. Hu Jin Tao, the outgoing president, who had tried to restrain the excesses of the market, gave a stilted speech. Xi Jin Ping spent the whole session shifting impatiently in his chair like a guy in a hurry at the barber shop.

The main resolution passed without opposition. It ratified the decision to write Hu Jin Tao's 'theory of scientific development' into the party constitution, so that the official ideology of Chinese communism would become (take a deep breath): 'Marxism-Leninism-Mao Zedong Thought, Deng Xiaoping-Theory, the Important Theory of the Three Represents and the Theory of Scientific Development'.

Think of it as a layer cake of pro-market ideology. Deng opened up China to global capitalism and destroyed the social welfare system that its traditional working class had relied on; Jiang's theory of 'Three Represents' ordered the party to start representing the new bourgeoisie; while Hu's theory of scientific development said the party should do so while tolerating overt corruption less, and while reducing the number of factories that kill and maim people.

As the congress was brought to a close, as if to herald the expected boredom and opacity expected over the next few years, the band struck up the most turgid rendition of the 'Internationale' I have ever heard. When I joined in the singing, my minder, a journalist from the party's press, looked astonished. 'You are a Marxist!' he blurted out, and not in a sympathetic way.

We didn't know it then, as the last notes of the 'Internationale' echoed into the empty seats and polished wood of the deserted balcony above us, but Marxism in China was on its way back. Albeit, Marxism of a not very Marxist sort.

It is laughable that anybody should have to learn about, let alone analyse, the 'thought' of a man who has never subjected himself to a critical interview, let alone a free election. But Xi Jin Ping's 'thought', which was officially added to the party's constitution in 2017, matters to everybody on the planet. In the years since he took control Xi has solved the riddle China-watchers had been puzzling over for decades: in what form, and on what timescale, will China emerge as a geopolitical superpower to reshape the world, matching its economic size and strength?

Xi's answer is: as a heavily state-controlled form of capitalism, overtly committed to Marxism as an ideology, and a lot sooner than you expected.

Instead of allowing the usual horse-trading between different factions – known as 'collective leadership' – Xi quickly placed all the party's administrative organs under his own control. He launched an anti-corruption campaign that has targeted 1.3 million officials, including twenty-seven members of the Central Committee and a sitting member of the Politburo. Amid unconfirmed rumours of a coup attempt against Xi, two leading members of the military committee and sixty generals were sacked and the army placed under his close command.

To understand what Xi is trying to do you have to understand what China is: a state capitalist economy in transition to becoming a market economy; a rural economy in transition to becoming an urban one; and a low-value exporter which can no longer rely on cheap labour alone to become a technologically modern superpower.

During the transition to the market, the ruling party became a vehicle for corruption and self-enrichment to the extent that it repeatedly triggered existential protest movements. The most obvious was the Tiananmen Square uprising of 1989, but massive worker resistance that took place in the Chinese rustbelt in the 1990s also haunts the leadership.

Xi's number one problem is to maintain consent among the Chinese working class, now expanded by 250 million migrant workers, for a single-party communist government. That is because every one of those 250 million experiences in their everyday life what the China expert Minxin Pei calls 'collusive corruption'. This is a system of graft, judicial favouritism, arbitrary policing and organized crime, which everyone knows is going on, which implicates many ordinary people in its activities, and which is held together by big, powerful informal networks.

Such networks undermine consent because they cream off large amounts of money from the state sector, which in turn eats into the provision of public services; in the private sector, meanwhile, their existence leads to inefficiency and lack of investment. So Xi's attack on corruption is focused on two targets: the eventual destruction of these massive graft networks; and, in the meantime, to stop them siphoning off billions of dollars in untaxed and undeclared incomes, which flow out of China and into the global finance system.

Xi's project is a response to the West's ongoing crisis. By 2012, when Xi took power, it was clear that the West was in long-term stagnation, and that China needed to create its own domestic market faster than had been thought.

After Trump's victory, things changed again. In May 2017 Trump's then national security adviser, H. R. McMaster, co-authored with economic adviser Gary Cohn an outline of Trump's foreign policy. 'The world,' they wrote, 'is not a "global community" but an arena where nations, nongovernmental actors and businesses engage and compete for advantage.'[1] No one understood the meaning of that better than Xi, who had been advocating the concept of a 'community of common destiny' – i.e. a new multilateral order – ever since Russia cut loose from the unipolar order back in 2008.

By January 2017, when Xi addressed the World Economic Forum

at Davos, the existential conditions for the Chinese elite's strategy had changed. America was no longer anything like a stable democracy. It was ruled by a crank determined to wage a trade war against Beijing; and America's willingness to guarantee the rules-based global order looked shaky.

With hindsight, therefore, Xi's launch of the so-called 'Belt and Road Initiative' in 2013 looks like genius. The 'belt' is a series of land infrastructure projects linking China with Europe via Central Asia; the 'road' is maritime, ports and related communications infrastructure linking the Chinese coastal economy to its client states Iran and Pakistan, and to the Suez Canal.

Belt and Road is often seen as a geopolitical gambit to make trading partners dependent on Beijing for development. But it is also an attempt to capture what remains of the 'catch-up growth' expected in the northern hemisphere. With its sparse population and massive natural resources, Central Asia is the last frontier of technological modernity. Turning it into a thriving corridor instead of an economic desert would guarantee a market for China's heavy industry long after its domestic market is sated, and reduce China's reliance on trans-Pacific trade routes.

It's impossible to predict if Xi's project will survive. It is so clearly outside the comfort zone of the mafia networks who have colonized the party that many China experts expect a backlash. However, if it does survive and bear fruit, it will add one more strange ingredient to the global ideological cocktail of the mid-twenty-first century: the bureaucratic, anti-humanist form of Marxism that died with the Soviet Union in 1991 will return, with massive prestige.

It is for this reason that developing our mental defences against Chinese official Marxism has to be among the reflexes of a radical humanist movement today.

Even within the confines of China's surveillance state, information technology has massively expanded people's ability to communicate and organize. In the 1990s, Western liberalism assumed that a Chinese middle class would simply emerge and democratize the CCP; indeed, there were signs of this under Hu Jin Tao, when 'constitutionalism' and 'universal rights' became codewords for experiments to separate the executive, the judiciary and China's token legislatures.

All this has been reversed under Xi. But the problem remains. A technologically modern population is going to demand greater freedom, and the only way to stop them is to impose inhuman levels of surveillance and control, which is exactly what Xi is doing.

In 2013 the CCP issued a directive known as 'Document 9', ordering teachers and academics to stop promoting seven ideas. These were: Western-style constitutional democracy; 'universal values'; civil society – which, the document says, is a Western concept being used to attack the party's legitimacy; a free market-oriented economic policy (in other words, neoliberalism); press freedom; and 'historical nihilism' – which is to say, the idea that the CCP's overall impact on China has been negative.

The seventh and final 'don't speak' concerned the use of Marxist terminology to criticize Xi within the party. Members are not allowed to say that China has become 'state capitalist' or that the state is a 'new bureaucratic form of capitalism'; nor are they allowed to call for faster political reform.[2]

The real danger of the 'don't speaks' can only be understood once you realize how Xi intends to utilize Marxism: as an all-powerful anti-humanist doctrine justifying one-party rule and algorithmic control.

The backlash against humanism under Xi is decisive. In 2012, the year he came to power, researchers studying the official Chinese media found 150 articles portraying the term 'universal values', 78 per cent of which viewed the idea positively. By 2013, out of 500 articles about 'universal values', 84 per cent were negative.[3] The same Orwellian reversal had happened with the term 'constitutionalism', which was attacked in no fewer than 1,000 headlines in that year.[4]

Nobody who understands the history of injustice perpetrated against China by the former colonial powers can object to its determination to remain geopolitically strong and independent. After Obama's 2012 pivot and the election of Trump it was inevitable that China's diplomatic stance should begin to reflect the emergence of multi-polarity and disorder. Xi has called for Chinese leadership in the critical technologies of semiconductors, artificial intelligence and biotech; that means head-to-head competition with both the USA and Japan. Combine this with the Belt and Road Initiative, and a clear path emerges for China to become a hemispheric superpower around 2030 and the global leader in technological innovation.

The problem is that at the same time it will become a major global supplier of anti-humanist ideas. Xi intends to fuse Marxism, Chinese nationalism and a paternalist form of Confucianism into a state ideology that will last throughout its journey to superpower status.

Marxism will be used to justify the continued purging of mafia networks; the suppression of rival factions in the party; and the total absence of any form of democratic participation or expanded human rights. That's what Xi meant when he told party leaders in 2017: 'If we deviate from or abandon Marxism, our party would lose its soul and direction. On the fundamental issue of upholding the guiding role of Marxism, we must maintain unswerving resolve, never wavering at any time or under any circumstances.'[5]

But the Marxism that is to be crammed down the throats of China's rising generation is the opposite of the real thing.

The real thing entered China via the writings of Chen Duxiu, a teacher, who in 1915 founded the magazine *New Youth*. As China's intelligentsia encountered Western thought in the 1890s, they had tried to marry it to the rigid, patriarchal Confucian ideology of the Qing dynasty under the slogan: 'Chinese knowledge for substance, Western knowledge for practice'.

Chen Duxiu's generation broke dramatically with this, urging people to abandon Confucianism, embrace scientific thought, fight imperialism, write fiction in the language of the people and to accept Western humanism.

From there it was a short journey to fully fledged Marxism. After the May Fourth Movement, an uprising against the imperialist powers in Shanghai in 1919, Chen founded the party that today rules China. Sidelined by Stalin, and declared persona non grata for criticizing Mao, Chen Duxiu is the humanist 'spectre' haunting Chinese Marxism.

In the decade before Tiananmen, intellectuals in China were allowed to re-engage – belatedly – with the humanist form of Marxism outlined in this book. Wang Ruoshui, then deputy editor of the *People's Daily*, published an article entitled 'Human beings are the starting point of Marxism', which argued that alienation existed even in planned economies like post-Mao China.[6]

It was not hard-line Maoism, however, which killed resurgent Chinese humanism. It was the pro-market faction around Deng Xiao Ping. In 1983 Deng made a speech condemning attempts to humanize Marxism as 'mental pollution' – and rejecting any idea that promoting human rights or universal values was part of the market reform process he had unleashed.[7] Wang was purged from the party in the mid-1980s, while other humanists were labelled 'thought criminals' for their support for the Tiananmen protest. The Chinese journey to the market was designed to produce a form of capitalism without respect for the individual human being.

According to Xi, Marxism is an anti-humanist doctrine of predestination, which says that any feelings of alienation, sadness or frustration felt by millions of Chinese people are illusory, and which justifies state control of both behaviour and thought.

Like his predecessors, Xi is determined to fuse this monolithic Marxism with a state-backed form of Confucianism. Confucius, like Aristotle, centred his concept of human goodness around an ordered society in which sons obeyed their fathers, women obeyed men and slaves obeyed their masters. But over two millennia, Confucian ideology and statecraft became a justification for inhumanity. By the end, they justified not just the absolute power and brutality of emperors, which you sense if you visit a place like the Forbidden City; they came to justify treating women as subhuman, and to explain why the poor could not resist oppression.

Technological modernity either enhances human creativity and control or is used to suppress it. But as the possibilities of algorithmic control, surveillance and artificial intelligence become obvious, Xi's China has developed the world's most comprehensive master plan for anti-humanism.

The planned Social Credit System, due for rollout in the 2020s, is the logical result. The Chinese state intends to force every citizen and every company into a common, state-run 'rating' system, in which everything from creditworthiness to political loyalty can be judged – not just by the state itself but by your peers. 'If trust is broken in one place, restrictions are imposed everywhere,' says a design document published in 2016. People who are rated untrustworthy will have reduced access to everything from the internet to foreign travel to bank loans to certain jobs.[8]

If you add to this the massive citizen database the Chinese state will accumulate, you have the beginnings of the first technologically empowered totalitarianism of the twenty-first century.

Xi Jin Ping has defined with the utmost clarity what the Chinese state must do to complete the modernization of the country without freedom and democracy. It has to suppress humanism in the name of Marx, and deploy algorithmic control on a scale unknown in any other country. If it works, you can be sure this kind of 'Marxism' will become highly popular in the West.

That's why rejecting the 'thought' of Xi Jin Ping is no side issue. China as much as the West needs a radical defence of the human being. It is comforting to know that this is what the CCP's leaders fear the most.

19

Never Give In

Nothing is signposted on the Ducos Peninsula; nobody is commemorated. You drive through an industrial estate, diesel fumes heavy on the air amid the ragged palm trees. The nickel plant, which dominates the bay, spews its yellow clouds into the Pacific sky. The mud crabs are still fighting though, just as in the memoirs of Louise Michel. As the sun falls towards the ocean, this outcrop on New Caledonia is as remote and bleak as she found it on her arrival here in 1873.

Michel was sentenced to deportation for life to the French Pacific colony, 1,000 miles east of Australia, for the crime of armed insurrection. The place where they isolated her, together with a handful of other female revolutionaries, is now called the Baie des Dames. You can still see where their grass and mud huts stood, huddled in the neck of an isthmus between two beaches. All that's left is an earth platform and some ornamental trees.

The place that once housed the most dangerous women in Europe has been obliterated by a Total oil depot, whose security guard spots me on their camera system and demands to know why I am taking pictures. When I answer 'Louise Michel', he points to the patch of levelled ground and says: 'She had to be segregated. Her views were extreme.'

He is correct in that. Louise Michel was an extreme advocate of human freedom and an extreme opponent of all attempts to categorize, control and limit what people can achieve.

Michel, a schoolteacher from the Paris slum of Montmartre, was sent to New Caledonia for instigating the first modern urban revolution: the Paris Commune of 1871. Within eighteen months of landing on the island she had transcribed and published the first ever collection of the legends and epic poems of the Kanak people, the indigenous Melanesian population of the island, who were regarded by the French as subhuman. In 1878, when the Kanaks staged an armed revolt against the whites, Michel was one of the few to support them. She claims to have given her red scarf, official regalia she had preserved from the Commune, to two warriors as they prepared to join the fighting:

> They slipped into the ocean. The sea was bad, and they may never have arrived across the bay, or perhaps they were killed in the fighting. I never saw either of them again, and I don't know which of the two deaths took them, but they were brave with the bravery that black and white both have.[1]

Most of Michel's biographers believe this story to be untrustworthy. The Kanak insurrection took place fifty kilometres to the north; the deportees were isolated and forbidden all contact with the indigenous inhabitants. If Louise Michel had met any Kanaks, said a man who'd spent his life collecting her memorabilia, they would most probably have killed her. But I came to this bay because I had a hunch that they were all wrong.

I can see from where I am standing that it would be possible to swim from here to the mainland, by hopping between small islands and rock shoals not shown on today's maps but marked on those of the time. By searching the archives of the revolt, I find that right there on the opposite shore, maybe two kilometres from where I'm standing, an entire clan of Kanaks was massacred on 29 June 1878 because an armed group of terrified white colonists thought they were about to join the insurrection.[2]

Everything else in this landscape is as Louise describes it. The *naiouli* trees still twist in the wind; the volcanic rocks stand like tombstones; the sea laps the mud of the shore and the crabs fight each other. Why would she lie about the Kanaks and the scarf? Why did sensible twenty-first-century scholars doubt that a white woman who opposed colonialism would support a rising of black people against her own government?

The answer is entirely relevant to our predicament: if you've never seen a revolution, you don't know what it can do to human beings. It's like coming across a beach strewn with wreckage, uprooted trees and dead seabirds, but never having seen the cyclone that left them there.

The crisis of globalization means that we, too, will see revolutions – unless the twenty-first century defies every observable pattern in history. The mismatch between the technologies now available and the economic forms they are trapped within has become too great not to produce upheaval. If you think talk of a revolution is far-fetched then you have to imagine someone like Xi Jin Ping one day granting multi-party democracy to China; or a Russian liberal government peacefully replacing Putin's party of crooks and thieves; or Bannon, Breitbart and the *Daily Stormer* simply closing down operations for lack of interest once Trump is impeached or indicted.

If so, we have to learn what revolutions mean. Yet, as the unmarked site at the Baie des Dames suggests, the stories of past revolutions are systematically walled off to us, impenetrable, portrayed as so different from our lives as to be irrelevant. Today, the peninsula is home to low-paid migrants from other Pacific islands; they inhabit shacks similar to the one Michel lived in, but most have probably never heard of her nor the immense event she was part of.

Unlike most political prisoners, Louise Michel actually committed the crime that she was sentenced for. She started a revolution. Before dawn on 18 March 1871 she ran through the streets of Montmartre, armed with a carbine, and summoned a crowd of women. By sunrise they had caused a mutiny in the French army, by noon they had got out of control, killed two generals and (in true Parisian style) cut up their horses for food. By nightfall they had begun the first experiment in working-class self-government in history.

The Paris Commune was the product of accidents and mistakes. In 1870 the emperor of France accidentally started a war with Germany. He accidentally got himself taken prisoner, turning France into an accidental republic. When the republican government surrendered Paris to the Germans, and were offered the choice to disband most of the army or the National Guard, they mistakenly chose to save the latter – which consisted of 100,000 armed workers, led by an unofficial central committee of republicans, communists, anarchists and socialists. When the people took to the streets to prevent the National Guard from being disarmed, the army's commanders mistakenly fled the city, leaving it in the hands of the National Guard, which then called elections for a democratic city council – the Commune.

On the second day of this concatenation of mistakes and accidents, as activists debated strategy in the occupied city hall of Paris, the artist Daniel Urrabieta Vierge was so taken by a mysterious woman dressed in a man's uniform, standing guard with a rifle and fixed bayonet, that he immortalized her in a sketch. That is Louise.

From 18 March to 28 May 1871 Paris was run by its people. The main political forces inside the Commune were ultra-left republicans, moderate anarchists and followers of Karl Marx. Louise Michel was part of the first group. A schoolteacher who taught herself to use a rifle at funfairs, she spent her time during the Commune organizing women to form ambulance brigades, fighting on the walls of Paris and speaking at the nightly meetings of revolutionary clubs in occupied churches.

When the French army retook Paris, they killed 30,000 people suspected of supporting the Commune, most of them non-combatants. Though students will rarely be taught this, the ghosts of those 30,000 haunt the most famous images in art.

When you look at any Impressionist painting of Paris produced in the 1870s, it is always worth asking: 'where are the missing people?' Does the artist want us to understand that those cobbled streets had recently been torn up to build barricades? That the bars and cabarets had only recently played host to a workers' government? In a painting like Gustave Caillebotte's *Paris Street, Rainy Weather* (1877), which shows a busy street full of well-off couples under umbrellas, why are their faces so anxious; and why are there no working-class people in the painting? When it was first exhibited, reviewers were very enthusiastic about the shiny wet cobblestones, which they noticed had been 'washed clean' by the rain.[3] Washed clean of what?

After the Commune was destroyed, the elites of Europe wanted to forget about it. That's why they sent 9,000 political prisoners from the most civilized city on earth to live on an island populated by

45,000 hunter-gatherers. But the working class could not forget. Because this was the first revolution in history where one form of oppression did not replace another.

The National Guard – a de facto working-class militia with battalions based in small neighbourhoods – ran itself. The revolutionary clubs Michel spoke in, though they were chaotic and ad hoc, took decisions: they passed resolutions, sent delegations, organized food distribution, terrorized potential traitors. In Michel's school and dozens like it they resolved to teach 'only ideas that have been scientifically proven' – that is, all the ideas not taught in the Catholic-dominated schools of Imperial France. The labour department of the Commune banned fines at work, passed rudimentary workplace regulations and ordered all factories abandoned by their owners to be turned into worker-run cooperatives.

The Paris Commune was, in short, the first ever semi-state. Marx – who'd sent a 23-year-old Russian woman called Elisabeth Dmitrieff to Paris as his emissary to try and radicalize things – didn't advocate that outcome. But he immediately understood it. Two days after the Commune was defeated, Marx penned an influential account of what had made it different.

The Commune proved, Marx said, that workers can't just take hold of the capitalist state but have to form a new kind of power. The delegates to the Commune were paid the same wages as a worker; they were instantly recallable to the localities they were elected by; they were overwhelmingly working class; there was no standing army and no political police force; and judges too were elected and recallable. Unlike all other revolutions, the Commune was 'the political form at last discovered under which to work out the economical emancipation of labour'.[4]

The Commune was massacred because – to the frustration of advocates of revolutionary violence – it was not violent enough. Its military offensives against the French army melted away; once the French army invaded Paris, resistance was strong only where guardsmen fighting on a barricade were surrounded by their wives, lovers, children, parents and neighbours. It drew its strength from the living spirit of a people suddenly freed from the standard bullshit of their time: from hierarchy, deference, the stigma of servants' uniforms and military ranks; from the disgrace meted out to prostitutes and the sanctity accorded to priests.

What Marx – and subsequent Marxists – never properly understood, however, was that the real revolution took place in people's lives. Instead of blowing the lid off the state, it was more like blowing the lid off society. And in that the Commune resembles more closely the short-lived and accidental uprisings of our century than the classic revolutions of the twentieth. In Greece, among the tent camps and the tear gas, I could feel the spirit of the Commune – not least because I saw one protester with Louise's face tattooed on her arm.

So what will our revolutions look like? Are we destined to wander dreamlike, as Michel did, into a revolutionary situation and then emerge once again defeated – with the kind of traumatized, poetic but damaged personality she was clinging to as they dumped her on a Pacific island?

To answer that we need to understand the accident the neoliberal elite fears most: the election of democratic left governments followed by the creation of truly democratic and transparent states.

Under neoliberalism, the state has become crucial to generating and distributing profits – through continual privatization, outsourcing and the coercive imposition of market values and metrics into ordinary life. If you watch the reaction of the elites to Jeremy Corbyn, Bernie Sanders, Alexandria Ocasio-Cortez and France's Jean-Luc Mélenchon, to Podemos in Spain and Syriza in Greece, the fear is always at its highest, the sabotage at its most blatant, wherever the left threatens to switch off the great privatization machine and start collecting tax on wealth and corporate profits. That's why, as punishment for resisting financial logic, the European Union and IMF decreed Greece had to privatize €50 billion-worth of public property.

Equally terrifying for the super-rich and the bankers is the idea of a state that suddenly dispenses with deep surveillance, nuclear weaponry and militarized policing and just lets ordinary democracy and an uncorrupted judiciary try to run things. This, far from being a utopian fantasy, was the prospect raised by two progressive national independence struggles in this decade: in Scotland and Catalonia.

On 1 October 2017 I witnessed the beginnings of a peaceful uprising in a major city. This was Barcelona, on the day its people voted for independence from Spain. On our mobile phones we could see the riot cops deployed from regions hostile to independence dragging pensioners by the hair, stamping on the fingers of young women, batoning the motionless bodies of old men, leaping down stairs to smash both boots into the ribcages of unresisting voters.

These images had been going viral from first light, as we milled around a polling station in the working-class suburb of Sant Andreu. Everybody in that throng, maybe 500 of us, expected to be on the receiving end of extreme police violence within hours. Against fear, the Catalans deployed mundane efficiency. They made people queue to vote. They refused to take votes while the 4G network was down, because the paper votes couldn't be verified without it. They let the elderly in first and made the young stand in the rain. As the elderly voters – many of whom had survived fascism – emerged they sang: 'we've already voted' and the crowd politely applauded.

Thousands of young people surrounded another polling station, the old university building of the Scuola Industrial, ready to defend it. I watched successive groups of youths trying to build a barricade. Like everyone who resists the neoliberal order in Europe, they knew what they were up against. The threat of economic violence is always the first defence line: companies will threaten to disinvest, central banks will pull support for the currency. Precisely these threats had been used to destroy Greece and stifle Scottish independence; the same threats were now deployed against Catalonia in the event that people voted to secede from Spain. But if economic violence doesn't work, there's always actual violence. Whatever the European Convention on Human Rights says, most people with a smartphone have worked this out for themselves.

All that day the Spanish police were jamming the phone network and randomly seizing ballot boxes. But the referendum went ahead anyway. Amid the baton slaps and rubber bullets, two million people managed to cast a countable vote, with 90 per cent voting Yes to independence.

In the El Clot district, a working-class suburb whose dense street pattern echoes its medieval origins, the voting stations were so close that, as one line of 1,000 people snaked around the block waiting to vote, another line formed across the street for a different polling station. And the real democracy wasn't just happening in the voting booths: it was happening in the few yards between the two queues.

Everybody stood in the rain and talked in small groups – quietly, civilly – about what to do. This street space, with its tobacco and occasional marijuana fumes, wet dogs and irascible pensioners, was alive with democratic argument. To save bandwidth, frantic officials marched up and down the queues of voters demanding everyone switch their phone to airplane mode – and most people complied.

Though two million voted Yes, most of those who opposed them simply stayed at home to try to weaken the legitimacy of the vote. But you have to weigh the quantity of democracy against its quality.

Alex, an eighteen-year-old law student whom I met trying to build the barricade at the Escola Industrial told me that for him this was not about flags and even language. Rather, he saw a Catalan-sized state, free of control by the Spanish financial elite, as the best way of protecting and enlarging his human rights. 'Drets humans, drets humans', a phrase I heard buzzing through countless conversations that day. By going peacefully onto the streets for hours, and creating in their own districts a real democracy of coexistence, toleration and pacifism, Catalans showed that the quality of their democratic culture was an order of magnitude greater than that of the closet fascists and Catholic reactionaries who pull the strings in Madrid.

The Catalan revolt and the Scottish independence campaign of 2014 raised a prospect much more radical than mere secession. In both countries, the supporters of independence understood that if you start a new state from scratch – even if you keep the economy just as capitalist as it was before – you are suddenly in a place where the elite has lost its power to lie to you, to cover up corruption, to bombard you with surveillance and to subject you to arbitrary repression.

This reveals an interesting fact about the modern state. If it had to be founded anew, conforming to modern concepts of human rights and accountability, it would lose large parts of its repressive apparatus. As a result, revolts that take the form of secession – by a city or a region – are more terrifying to authoritarians than outright attempts to take the entire state at once.

The realities of elite control are always based on decrepit things: in

Spain's case, the monarchy, the deep state, the business corruption networks and the militarized riot police. So at one level, all revolts against neoliberalism simply call its bluff. They ask – since the market and individual choice are supposed to be paramount – why do we need a repressive state to dictate, limit and control our choices? Likewise, the basic form of all neoliberal counter-revolutions is the imposition of militarized policing, arbitrary justice and media control.

That's why, as punishment merely for staging a referendum in defiance of the constitution, the Spanish state used its Supreme Court to arraign and detain the mainstream Catalan political leadership and put them into solitary confinement, mobilizing fascist groups to intimidate their followers, while Madrid's riot police looked on passively.

There is no ideal form of revolutionary government, but the one we know elites fear the most consists of street-level democracy – continuous face-to-face discussions and debates – plus non-hierarchical decision-making groups in continuous session. This is what the Paris Commune achieved. And it is what lies within reach wherever the population of a major city decides it has had enough of elite control.

It should be clear now what a revolution is: the temporary achievement of true human status, a glimpse of what Marx called 'species being'. When we compress years of change into a few days or weeks we also speed up what Marx called 'the alteration of humans on a mass scale'. We begin to live for each other.

By the same logic you can understand what a counter-revolution is. It is the re-imposition of selfishness and anti-human routine. The aim of counter-revolutions is to eradicate the memory of our experience of human self-transformation in the moment of revolt. That is why mainstream historians have made a special effort to prove that the detachments of female fighters Louise Michel organized were a myth.[5]

If you want to know how far counter-revolutions will go, visit La Foa, the rural town in New Caledonia where, in 1878, the indigenous Kanak people began their rebellion against France. At the Hotel Banu, a sleepy old French café, the walls are covered with memorabilia of deer hunting and old photographs of the white settlers who made this place their home. It's a touching memorial to history and tradition – but there's something missing. There are no photos of Kanaks on the wall. Nor, today, are there many indigenous people on the roadside or working in the fields, such as you would find in most developing countries. In fact there are not many people at all. An island the size of Portugal has just 279,000 inhabitants.

Yvan Kona, a Kanak historian, told me how his people's history had been expunged. After the rising of 1878 the region was systematically depopulated by the French. By the late nineteenth century an island population of 45,000 had been reduced to 16,000. Even today there are only 100,000 Kanaks in New Caledonia – outnumbered by white colonial-era settlers, French civil servants and migrant workers from other islands. But after the Kanak people staged their third big revolt in the mid-1980s, the French state finally began treating them with respect, allocating money for jobs and education, and creating a parallel Kanak ruling council on the island.

However, despite the extreme sensitivity of the Kanak question – with independence from France always on the table – there is, Kona told me, total insensitivity towards the preservation of their story:

> If you look at the land, on the crests of the hills and down in the valleys, you will see isolated pines and coconut trees. These trees were tribal symbols showing that there was a tribe that lived there, that these were ancient lands. The colonialists began by destroying these signs. First they destroyed these signs. They destroyed us.

Though his work involves gathering oral records from the surviving clans and tracing the locations of their tribal gardens and villages, Kona tells me, he is regularly refused access to the land by white landowners. In this way, the story of the Kanaks is still being suppressed.

For all her naivety, her Eurocentricity and her mistakes Louise Michel understood what it means never to give in. Moved from her urban element to a jungle, she asked what the power struggle of that jungle was, and plunged into it. She understood the revolutionary importance of preserving and telling stories from the viewpoint of the vanquished and so should we.

20

Live the Antifascist Life

In 1977, the sociologist Michel Foucault penned a half-serious attempt to write an ethical code for the postmodern era. Satirizing a famous Catholic manual on morals, he called it an *Introduction to the Non-fascist Life*. The fascism we need to resist, wrote Foucault, is not just the fascism of the far right. It is 'the fascism in us all, in our heads and in our everyday behaviour, the fascism that causes us to love power, to desire the very thing that dominates and exploits us'.[1]

As with the seven Christian virtues, Foucault offered seven rules. Do not seek power. Do not embrace an over-arching political goal. Reject hierarchies. Reject the idea that negativity is politically effective. You don't have to be sad in order to be a political activist. Do not try to base political action on claims about truth. And do not base politics on human rights, or indeed individual human beings. 'The group,' wrote Foucault, 'must not be the organic bond uniting hierarchized individuals, but a constant generator of de-individualization.'

Though few people have read the actual text, it is fair to say that Foucault's commandments have been widely adopted among those trying to resist globalization, climate change and repressive states using the methods of horizontalism. A whole generation of activists has attempted to dissolve power through networked activism, by shape-shifting from one kind of struggle to the next, and through the strategy of 'one No, many Yeses'.

The logical problem for Foucault was: why write an ethical system for individuals when you are advocating they dissolve and shatter their separate selves? If 'man' is a recent invention and about to disappear, why bother yourself with a set of rules for him to follow?

However, this problem did not stop Foucault gravitating even more strongly towards a form of virtue ethics in the final years of his life. In an interview in 1984, he explained how the ethical practice of ancient Greek and Roman slave-owners became so focused on 'care of the self' that they turned their own lives into a work of art and, by restraining their desires, stopped oppressing other (free) people.[2] When asked whether this practice of self-care could become the basis of a new philosophy for the present or a political alternative, Foucault answered that he had 'not investigated the question'.

But it is obvious with hindsight that 'self-care' and 'turning your life into a work of art' have become a whole new religion among the middle class of the developed world. From the gym to the yoga mat to the operating table at the plastic surgeon's, entire industries have sprung up around 'care of the self'.

The problem is that, as with the Greeks, it has not eradicated inequality and injustice. More to the point, it has not stopped the rise of a modern neofascism. If you want to study today's equivalents of the Greek aristocrats who were good at 'self-care', the millionaire alt-right ideologist Milo Yiannopoulos would be a legitimate place to start. Or Marine Le Pen. Or the neat, hipster Identitarians patrolling the Austrian border, carrying flags signalling that, like the Spartans, they will repel dark-skinned invaders. Or Trump himself.

Foucault was right to state that, as the revolutionary wave of 1968 subsided, its failure was the failure of a hierarchical politics based on power: male trade unionism, Stalinism, the doomed urban guerrilla movements of the Black Panthers and the Red Brigades. He was also right to imply that left-wing totalitarianism in the twentieth century had derived much of its power from the old Christian instruction to be 'self-less'.

But his techniques for living a 'non-fascist life' do not solve the problems facing us today. We need, for the reasons I have explored in this book, to risk gaining conventional political power. We need to engage with the state – militarized and oppressive as it is – and the electoral system, because unless we do so the forces of liberal centrism will collapse, to be replaced with authoritarian nationalism.

The generation that has absorbed Foucault's ethical principles needs to move beyond them. It is, of course, possible to live a 'non-fascist life' – even in a place like Arizona, where the cops snatch migrants off the streets, where inhuman jail conditions are promoted as a deterrent, and where white supremacists set the Republican agenda. You could sail from one bubble to another: from the pilates gym to the shrink, attending anti-Trump protests in Washington and donating to the Democrats. But such behaviour doesn't win power.

Instead, we are going to have to learn afresh what it means to live an *anti*fascist life.

During the Spanish Civil War, George Orwell met an Italian anarchist fighting in the militia of a far-left party called the POUM. Orwell would write about this meeting numerous times – in his book *Homage to Catalonia*, in a poem called 'The Italian Soldier' and in a bitter essay written during the darkest days of the Second World War.

He described the Italian soldier as a semi-literate peasant, unable to read a map but perfectly capable of knowing which side of history he was on. They immediately liked each other. Orwell thought he was a man able both to 'to commit murder and to throw away his life for a friend'.[3] By the middle of the Second World War, Orwell wrote, as a far-left opponent of both fascism and Stalinism the soldier was certain to be dead: 'He symbolizes for me the flower of the European working class, harried by the police of all countries, the people who fill the mass graves of the Spanish battlefields and are now, to the tune of several millions, rotting in forced-labour camps.'[4]

But history is composed of real people. Assuming he was not some romanticized, composite character created by Orwell's imagination, what more could we know about the Italian soldier?

Orwell joined the POUM militia in Barcelona in December 1936. By then a majority of the Italian anarchists and left-wing communists who'd signed up to fight that summer had quit over the militia's imposition of military discipline. Of those who remained – numbering at most twenty-five – only Cristofano Salvini had the 'reddish-yellow hair and powerful shoulders' Orwell mentions in *Homage to Catalonia*. Salvini, then, is the prime candidate to be the Italian soldier.[5] Let me tell you his story.

Cristofano Salvini was born in 1895 in Casole d'Elsa, an ancient rural town southwest of Florence, where he worked as a bricklayer.

In 1920, the year after the Italian Socialist Party became the biggest party in parliament, he became its local councillor. The following year he was part of the split from the socialists that formed the Italian Communist Party. In 1923, he fled Mussolini's fascist government for France, where he joined one of the many Trotskyist groups there. In August 1936, at the first call for volunteer fighters, Salvini went to Spain as part of a group of fifty people called the International Lenin Column, and was thrown immediately into front-line combat at Huesca.

But it turns out that, whatever his deficiencies at map reading, Salvini was highly literate. Over the next eighteen months he authored two extended polemics in French newspapers denouncing the Stalinist takeover of the Republican side in the Spanish Civil War. The internal political debates of the unit he fought with were documented meticulously in the POUM's weekly newspaper. When, in May 1937, the Spanish communists moved to suppress anarchist and far-left groups in Barcelona, Salvini disappeared and was presumed dead. In fact, he had joined an anarchist-run militia unit on a different front.

After the Civil War he escaped to France, and was interned in a labour camp. In 1940 he was captured by the Germans at Dunkirk and repatriated to Italy, where he was sentenced to five years in prison. Released when left-wing partisans liberated Tuscany in 1943, he went back to his old job as a bricklayer in Casole d'Elsa and died in 1953.[6]

He had done all the things Foucault tells us to avoid. He had believed in truth, fought for a totalizing project and taken very little care of the 'self'. Yet he had lived an antifascist life.

'But the thing that I saw in your face, no power can disinherit,' wrote Orwell, commemorating people like Salvini. 'No bomb that ever burst shatters the crystal spirit.'[7] But how do you create the crystal spirit in the first place? What made a bricklayer from a town organized around a single medieval street spend his entire life enmeshed in organizational politics and Marxist theory, in an effort to overthrow capitalism?

The answer is: it took two generations. In 1892 the disparate strands of the Italian workers' movement came together to form a socialist party. It survived initial attempts at repression and in the first fifteen years of the twentieth century grew its vote to 34 per cent.

After the First World War, a time when Italy experienced rapid indus-
trialization, Italian workers seized their factories and tried to run
them under workers' control. In 1921, the ruling class abandoned its
attachment to liberalism and backed the fascist movement led by former socialist Benito Mussolini. One account of how the fascists
operated in places like Casole d'Elsa says:

> In small towns, where everyone knew everyone, fascists inflicted rit-
> ual humiliation on their enemies, a powerful strategy of terror
> understood by all. Blackshirts forced their opponents to drink castor
> oil and other purgatives, and then sent them home, wrenching with
> pain and covered in their own faeces . . .They also accosted their
> opponents in public, stripped them naked, beat them, and handcuffed
> them to posts in piazzas and along major roadways.[8]

The targets for this violence were not members of an illegal group
or terrorists, but simply town councillors like Salvini, a bricklayer,
around whom an entire network of left-wing clubs, parties, trade
unions and cultural centres operated. These institutions were closed
down and the Italian labour movement torn up by its roots. That is
the reality of fascism.

Why did a modernist, technologically savvy and originally liberal
elite like the Italian ruling class suddenly need to do this? Because, in
the space of three decades, the Italian working class had moved from
identifying itself as a social force and defending itself, to working out
what it wanted to achieve – which was to replace capitalism with
socialism. They found each other, acted, and then identified an end
point for their actions.

People like Salvini lived through this process in a single lifetime –
moving from bricklaying to local politics, to strategic national
discussions about politics, to exile and ultimately armed conflict in a
militia unit named after Lenin. This experience was shared in different forms not by a few but in fact by hundreds of thousands. It's a
process Marx predicted, describing it memorably as the working
class becoming 'for itself'.

It is unlikely that – even in the struggles to come in countries such
as China, Bangladesh and Brazil – the industrial working class will
ever again achieve the level of social density that it did between 1900

and the late 1970s. The force that will have to resist fascism this time around is simply the atomized mass of people around you.

It includes everybody with an interest in reversing the commercialization of daily life, everybody sick of being coerced and 'nudged' into artificially competitive behaviour, everybody with a material interest in saving the planet from climate change, every woman who does not want to become the target of violent misogyny, every member of an ethnic minority who does not want to be depicted as an alien threat in the tabloid media.

We saw above that for Marx the proletariat had to go beyond common struggles to become a class 'for itself'. By analogy, the next phase for the more amorphous, less rigidly defined demographic of networked, freedom-loving people is to find out what it means to become collectively 'for themselves', just as Cristofano Salvini's generation did.

I don't claim that networked individuals, together, form a 'new class' in the Marxist sense. But they don't need to. As Edward Thompson pointed out in his study of eighteenth-century England, class struggles can take place even without clear and rigid social classes. Classes, said Thompson, are formed in struggle. People find themselves in a society structured around power and inequality, they experience exploitation and oppression, they identify points of common interest and they start to fight. The nineteenth- and by extension twentieth-century working class was, Thompson suggested, probably a special case in the clarity of its demarcations.[9]

At one level, the move from small-scale horizontal activism into national political parties such as Britain's Labour Party or the US Democrats is part of the evolution of networked individualism towards a common project of emancipation. But the next level has to go beyond a mere change in organizational forms – from the protest to the takeover of parties. It has to connect our individual refusal and defiance with a political project: to end market logic and promote human and environmental logic in its place.

In the 1920s, if a worker wanted to improve their situation they joined a union, waited patiently until everyone was ready to act and took goal-oriented actions, such as a strike or an electoral campaign. That's why people like Salvini were so focused on getting big organizations to change tactics – and presumably why he didn't quit the POUM militia when, after the first weeks of trying to fight without hierarchy, it appointed a few officers and allowed them to issue orders.

By contrast twenty-first-century capitalism is so complex that, even by taking individual or small-scale actions, we can have big effects. The left-wing writer John Holloway argues that the way to achieve what's needed is 'to create cracks in capitalist domination, spaces or moments in which we live out our dream of being human'.[10] Spaces where we act as if human values and not market values predominate.

That strategy was not alien to Cristofano Salvini's generation, but they grew beyond it, understanding that only highly organized formations of workers can defeat a highly organized capitalism of the kind they faced in Italy after 1922 and Spain in 1936, and achieve what Orwell described as a 'decent, fully human life'.

Today, however, we're at an impasse. The last thirty years have seen the ideological triumph of individualism, and then its failure. We've not yet found the courage to stop sacrificing to the modern equivalent of the Roman gods, even though we don't believe in them. And we rightly fear repeating the cruelty and injustice that were the result of twentieth-century people's commitment to single causes.

On Bondi Beach, Australia, where I am writing this, there are thousands of ordinary young people bathing in the sunlight. On the beach the competitive rituals they observe in the city are suppressed. You have great muscles? So do fifty people within a hundred metres of you. You can surf? Someone on the next towel can execute a perfect handstand. You have the latest iPhone? So does almost everyone else. You are special and unique, but so are most people – and modern popular culture implicitly acknowledges this.

On beaches, in the city squares at night-time and at summer dance festivals, the networked generation displays a profound human-ness and empathy that could easily be the foundation for a new social ethos. But leave the sand of Bondi and you're back in the world where 'self-care' is more important than solidarity. Bondi's main street is a parade of Lamborghinis and haute couture, its restaurant tables bookable only in two-hour slots. This is not really a class divide (though class still exists). It is a divide between performative capitalism and authentic human life. Or, if we are lucky, between the past and the future.

While he was still a Marxist, the philosopher Alasdair MacIntyre wrote that because class society suppresses human potential so thoroughly, 'human development takes place in quite unpredictable leaps. We never perhaps know how near we are to the next step forward.'[11] For me, all the angst emanating from misogynists, ethnic nationalists and authoritarians everywhere is evidence that they too can feel how close we are to that 'next step forward'.

Living the antifascist life involves putting your body in a place where it can actually stop fascism, and having done so, to hold a tiny piece of liberated space long enough for other people to find it, populate it and live. The radical defence of the human being starts with you.

Notes

1. Deutscher, Isaac, *The Prophet Outcast: Trotsky, 1929–1940*, London, 2003, p. 399

PART I: THE EVENTS

1. Arendt, Hannah, *The Origins of Totalitarianism*, London, Kindle edn, p. 435

1. DAY ZERO

1. https://yougov.co.uk/news/2016/11/16/trump-brexit-front-national-afd-branches-same-tree/
2. Human Rights Watch, https://www.hrw.org/world-report/2017/country-chapters/syria
3. http://www.bbc.co.uk/news/world-asia-china-34592186
4. Dendle, Peter, *The Zombie Movie Encyclopedia*, Jefferson, 2001
5. Weyl, Hermann, *Philosophy of Mathematics and Natural Science*, Princeton, 1950, pp. 65–6
6. Baudrillard, Jean, *The Illusion of the End*, Stanford, 1994
7. http://www.lemonde.fr/disparitions/article/2009/11/04/1979-on-m-a-souvent-reproche-d-etre-antihumaniste_1262644_3382.html
8. http://vhemt.org

2. A GENERAL THEORY OF TRUMP

1. https://www.theguardian.com/commentisfree/2016/nov/09/globalisation-dead-white-supremacy-trump-neoliberal

2. https://www.ft.com/content/adfaf156-39cb-11e5-8613-07d16aad2152?mhq5j=e1

3. http://time.com/3923128/donald-trump-announcement-speech/

4. https://poll.qu.edu/national/release-detail?ReleaseID=2264

5. http://www.democracycorps.com/In-the-News/a-new-formula-for-a-real-democratic-majority/

6. Johnston, David Cay, 'Just what were Trump's ties to the mob', *Politico* Magazine, 22 May 2016, http://www.politico.com/magazine/story/2016/05/donald-trump-2016-mob-organized-crime-213910

7. https://www.nytimes.com/2015/07/19/us/politics/trump-belittles-mccains-war-record.html

8. Fromm, Erich, *Escape from Freedom*, New York, 1941

9. https://www.cato-unbound.org/2009/04/13/peter-thiel/education-libertarian

10. https://www.theguardian.com/books/2017/jun/10/naomi-klein-now-fight-back-against-politics-fear-shock-doctrine-trump

11. Marx, Karl, 'Letter to Engels in Manchester', 30 April 1868, *MECW*, vol. 43, p. 20

12. Bellamy Foster, John, 'The financialization of capital and the crisis', *Monthly Review*, vol. 59, issue 11, 2008

13. http://www.cbpp.org/research/economy/chart-book-the-legacy-of-the-great-recession

14. Mayer, Jane, 'The reclusive hedge-fund tycoon behind the Trump presidency: How Robert Mercer exploited America's populist insurgency', *New Yorker*, 27 March 2017

15. https://www.bloomberg.com/news/articles/2016-11-21/how-renaissance-s-medallion-fund-became-finance-s-blackest-box

16. http://www.thedailybeast.com/libertarianism-30-koch-and-a-smile

17. Howe, Neil, 'Where did Steve Bannon get his worldview? From my book', *Washington Post*, 24 February 2017

18. https://www.wired.com/story/what-did-cambridge-analytica-really-do-for-trumps-campaign/

19. https://www.theguardian.com/commentisfree/2016/nov/09/globalisation-dead-white-supremacy-trump-neoliberal

20. Cox, Daniel, Lienesch, Rachel and Jones, Robert P., 'Beyond Economics: Fears of Cultural Displacement Pushed the White Working Class to Trump', *PRRI/The Atlantic Report*, 9 May 2017

21. Rothwell, Jonathan and Diego-Rosell, Pablo, 'Explaining Nationalist Political Views: The Case of Donald Trump', Draft Working Paper, November 2016, https://papers.ssrn.com/sol3/papers.cfm?abstract_id=2822059

22. Schaffner, Brian F. et al, 'Explaining White Polarization in the 2016 Vote for President: The Sobering Role of Racism and Sexism', Conference on the U.S. Elections of 2016: Domestic and International Aspects, IDC Herzliya Campus, 8–9 January 2017, http://people.umass.edu/ schaffne/schaffner_et_al_IDC_conference.pdf

23. https://www.washingtonpost.com/news/monkey-cage/wp/2017/12/15/ racial-resentment-is-why-41-percent-of-white-millennials-voted-for- trump-in-2016/?utm_term=.860ab7e418ba

24. http://www.foxnews.com/politics/2017/10/31/how-paul-manafort-is- connected-to-trump-russia-investigation.html

25. https://www.nbcnews.com/news/us-news/flynn-never-told-dia-russians- paid-him-say-officials-n756421

26. http://www.motherjones.com/politics/2017/09/whos-telling-the-truth- about-the-russia-meeting-kushner-or-trump-jr/#

27. Hoffman, Frank, *Conflict in the 21st Century: The Rise of Hybrid War*, Arlington: Potomac Institute for Policy Studies, 2007

28. https://www.theatlantic.com/international/archive/2014/02/the-syrian- opposition-is-disappearing-from-facebook/283562/

PART II: THE SELF

1. https://www.blaetter.de/archiv/jahrgaenge/2011/februar/kooperieren- oder-scheitern

3. CREATING THE NEOLIBERAL SELF

1. Davies's version reads: 'the elevation of market-based principles and techniques of evaluation to the level of state-endorsed norms'. Davies, William, *The Limits of Neoliberalism: Authority, Sovereignty and the Logic of Competition*, Sage Publications, 2016, Kindle edn, p. xiv

2. https://www.marxists.org/reference/subject/economics/keynes/general- theory/ch24.htm

3. For a full account see Mason, Paul, *Postcapitalism: A Guide to Our Future*, London, 2015

4. Sachs, Jeffrey and Wsyplosz, Charles, 'The economic consequences of President Mitterrand', *Economic Policy*, 1986, http://graduateinst itute.ch/files/live/sites/iheid/files/sites/international_economics/shared/ international_economics/prof_websites/wyplosz/Papers/Mitterrand% 20Sachs%20Wyplosz.pdf

5. Ibid., p. 290
6. Smith, W. Rand, *The Left's Dirty Job: The Politics of Industrial Restructuring in France and Spain*, Pittsburgh, 1998, p. 200
7. Arocena, Pablo, 'The Privatisation of the Public Enterprise Sector in Spain: Stuck Between Liberalisation and the Protection of National Interest', CESifo Working Paper No. 1187, May 2004
8. Boughton, James M., *Silent Revolution: The International Monetary Fund 1979–1989*, Washington, 2001, p. 320
9. Ibid., p. 237
10. Ganesh, G., *Privatisation Experience Around the World*, New Delhi, 1998, p. 203
11. Kim, Kwan S., 'Mexico: The Debt Crisis and Options for Development Strategy', Kellogg Institute Working Paper 82, September 1986
12. Attali, Jacques, *Verbatim I*, Paris, 1993, p. 399
13. Perlman, Janice, *Favela: Four Decades of Living on the Edge in Rio de Janeiro*, Oxford, 2010, p. xxi
14. See for example, Davis, Mike, *Planet of Slums*, London, 2006
15. Perlman, *Favela*, p. 107
16. Dahlman, Carl J., 'The rise of China: Implications for global growth and sustainability', in Fu, Xiaolan (ed.), *China's Role in Global Economic Recovery*, Abingdon, 2012, p. 108
17. Freeman, Richard, 'The great doubling: The challenge of the new global labor market', in Edwards, John et al (eds.), *Ending Poverty in America: How to Restore the American Dream*, New York, 2007
18. http://tass.com/economy/916534
19. https://wikileaks.org/gifiles/attach/144/144365_RussianoligarchPDF.pdf, p. 3
20. Frisby, Tanya, 'The rise of organised crime in Russia: Its roots and social significance', *Europe-Asia Studies,* vol. 50, no. 1, 1998, p. 31
21. Ibid.
22. Pelevin, Victor, *Babylon* (trans. Blomfield, Andrew), London, 2000, p. 9
23. Boltanski, Luc and Chiapello, Eve, *The New Spirit of Capitalism*, London, 2005
24. Sennett, Richard, *The Corrosion of Character: The Personal Consequences of Work in the New Capitalism*, New York, 1998
25. https://data.oecd.org/money/broad-money-m3.htm#indicator-chart
26. http://www.goldcore.com/us/gold-blog/global-debt-now-200-trillion/
27. https://fraser.stlouisfed.org/files/docs/publications/frbnyreview/pages/1990-1994/67192_1990-1994.pdf
28. http://ritholtz.com/wp-content/uploads/2011/03/nipao328111_big.gif

29. Walter, Per and Krause, Par, 'Hedge Funds – Trouble Makers', *Sviergesbank Quarterly Review*, 2000
30. Ravenhill, Mark, *Ravenhill Plays: 1: Shopping and F***ing; Faust is Dead; Handbag; Some Explicit Polaroids*, London, 2013, Kindle edn
31. http://www.mirror.co.uk/money/you-used-store-card-1990s-7114139
32. Lapavitsas, Costas, 'Financialised Capitalism: Crisis and Financial Expropriation', RMF Paper Number 1, February 2009
33. Ravenhill, Plays:1
34. https://www.federalreserve.gov/monetarypolicy/files/FOMC 19950201 meeting.pdf
35. https://www.theguardian.com/business/2003/apr/29/8

4. TELEGRAMS AND ANGER

1. https://www.marxists.org/reference/archive/hegel/works/1818/inaugural.htm
2. https://www.marxists.org/reference/archive/hegel/works/letters/1806-10-13.htm
3. https://www.marxists.org/reference/archive/hegel/works/pr/prstate.htm#PRa258
4. Hegel, G. F. W, *The Philosophy of History* (trans. Sibree, J.), Kitchener, 2001, p. 121
5. Hansard, *The Parliamentary Debates from the Year 1803 to the Present Time . . .*, vol. 32, 1 February to 6 March 1816, pp. 71–113
6. Fukuyama, Francis, 'The End of History?', *The National Interest*, Summer 1989
7. Ibid.
8. Ibid.
9. https://www.foreignaffairs.com/articles/1991-02-01/unipolar-moment
10. Greenspan, Alan, *The Age of Turbulence: Adventures in a New World*, New York, 2007
11. http://www.nytimes.com/2004/10/17/magazine/faith-certainty-and-the-presidency-of-george-w-bush.html
12. http://www.nytimes.com/2003/06/04/opinion/because-we-could.html?mcubz=0
13. http://researchbriefings.parliament.uk/ResearchBriefing/Summary/CBP-7877#fullreport and https://www.globalsecurity.org/military/world/china/budget-table.htm
14. Little, Bruce, 'Global Imbalances – Just How Dangerous?', *Bank of Canada Quarterly Review*, Spring 2006, p. 3

15. Brender, Anton and Pisani, Florence, *Global Imbalances and the Collapse of Globalised Finance*, Brussels, 2010, p. 2
16. http://www.nber.org/chapters/c3625.pdf
17. Foucault, Michel, *The Birth of Biopolitics*, Basingstoke, 2008, p. 207
18. Brown, Wendy, *Undoing the Demos: Neoliberalism's Stealth Revolution*, Brooklyn, 2015, Kindle Location 1103

5. THE CRACK UP

1. Alessandri, Piergiorgio and Haldane, Andrew, 'Banking on the State', Bank of England, November 2009
2. Haldane, Andrew, 'Rethinking the Financial Network', Speech, 28 April 2009, https://www.bis.org/review/r090505e.pdf
3. http://www.economist.com/node/16741043
4. Ibid.
5. Davies, William, *The Limits of Neoliberalism: Authority, Sovereignty and the Logic of Competition, SAGE* Publications, 2016, Kindle edn
6. http://www.bankofengland.co.uk/publications/Pages/speeches/2016/885.aspx
7. https://www.bl.uk/20th-century-literature/articles/howards-end-and-the-condition-of-england
8. See Mason, Paul, *Why It's Kicking off Everywhere: The New Global Revolutions*, London, 2012
9. Castells, Manuel, 'Materials for an exploratory theory of the network society', *British Journal of Sociology,* vol. 51, issue 1, January/March 2000, pp. 5–24
10. Castells, Manuel, *Networks of Outrage and Hope: Social Movements in the Internet Age*, Cambridge, 2012
11. Crabapple, Molly and Penny, Laurie, *Discordia*, London, 2013, Kindle Locations 917–20

6. THE ROAD TO KEKISTAN

1. http://www.nbc29.com/story/38204693/settlements-from-unite-the-right-05-16-2018
2. https://www.axios.com/what-steve-bannon-thinks-about-charlottesville-1513304895-7ee2c933-e6d5-4692-bc20-c1db88afe970.html
3. Georgiadoua, Vasiliki et al, 'Mapping the European far right in the 21st century: A meso-level analysis', *Electoral Studies*, vol. 54, August 2018, pp. 103–15

NOTES

NOTES

4. https://www.cato-unbound.org/2009/04/13/peter-thiel/education-libertarian

5. http://www.thedarkenlightenment.com/the-dark-enlightenment-by-nick-land/

6. http://www.bbc.co.uk/blogs/newsnight/paulmason/2009/09/g20_americas_struggle_to_adapt.html

7. http://crooksandliars.com/david-neiwert/beck-goes-nuts-over-hcr-concludes-ev

8. https://arstechnica.com/gaming/2014/09/new-chat-logs-show-how-4chan-users-pushed-gamergate-into-the-national-spotlight/

9. https://expandedramblings.com/index.php/4chan-statistics-facts/

10. https://www.forbes.com/sites/curtissilver/2018/01/09/pornhub-2017-year-in-review-insights-report-reveals-statistical-proof-we-love-porn/#345e1ae224f5

11. http://knowyourmeme.com/memes/red-pill

12. http://www.thedarkenlightenment.com/the-dark-enlightenment-by-nick-land/

13. https://www.independent.co.uk/news/world/europe/a-day-in-the-life-of-vladimir-putin-the-dictator-in-his-labyrinth-9629796.html

14. http://www.newsweek.com/dmitry-medvedevs-grand-strategic-ambitions-84943

15. https://foreignpolicy.com/2011/10/11/americas-pacific-century/

16. http://12mars.rsf.org/2014-en/2014/03/11/russia-repression-from-the-top-down/

17. https://www.amnesty.org/en/latest/news/2016/11/russia-four-years-of-putins-foreign-agents-law-to-shackle-and-silence-ngos/

18. https://warontherocks.com/2014/07/on-not-so-new-warfare-political-warfare-vs-hybrid-threats/

19. Thornton, Rod, 'The changing nature of modern warfare', *RUSI Journal*, 160:4, 2015, pp. 40–48

20. https://www.theatlantic.com/international/archive/2013/10/russias-online-comment-propaganda-army/280432/

21. https://globalvoices.org/2015/04/02/analyzing-kremlin-twitter-bots/

22. https://medium.com/@patrissemariecullorsbrignac/we-didn-t-start-a-movement-we-started-a-network-90f9b5717668

23. Olson, Joel, 'Whiteness and the polarization of American politics', *Political Research Quarterly*, vol. 61, no. 4, 2008, pp. 704–18

24. https://rollingout.com/2018/07/04/ranking-the-most-racist-acts-of-white-people-calling-the-police-on-black-folks/

25. https://en.wikipedia.org/wiki/Leigh_(UK_Parliament_constituency)#Elections_in_the_2010s

26. Fromm, Erich, *The Working Class in Weimar Germany: A Psychological and Social Study*, Leamington Spa, 1983, p. 43

27. Forscher, Patrick S. and Kteily, Nour, 'A psychological profile of the alt-right', *PsyArXiv*, January 2018

7. READING ARENDT IS NOT ENOUGH

1. Arendt, Hannah, *The Origins of Totalitarianism*, New York, 1958
2. Ibid.
3. Ibid.
4. Arendt, Hannah, *Eichmann in Jerusalem: A Report on the Banality of Evil*, London, 2006
5. Todorov, Tzvetan, *Hope and Memory: Lessons from the 20th Century*, Princeton, 2003, pp. 6, 314
6. http://www.timetableimages.com/ttimages/dlh4001i.htm
7. https://www.orwellfoundation.com/the-orwell-foundation/orwell/poetry/the-italian-soldier-shook-my-hand/
8. Arendt, Hannah, *Essays in Understanding: 1930–1953: Formation, Exile and Totalitarianism*, New York, 1994, p. 224
9. In van der Linden, Marcel, *Western Marxism and the Soviet Union: A Survey of Critical Theories and Debates Since 1917*, Leiden, 2007, p. 73
10. http://sovietinfo.tripod.com/ELM-Repression_Statistics.pdf
11. https://www.marxists.org/archive/rizzi/bureaucratisation/index.htm
12. Arendt, *Origins of Totalitarianism*, p. 159
13. Ibid., p. 435
14. Arendt, Hannah, 'Some questions of moral philosophy', *Social Research*, vol. 61, no. 4, Winter 1994
15. Nietzsche, Friedrich, *Complete Works,* Edinburgh and London, 1909, vol. 2, p. 325
16. Ibid., vol. 16, p. 184
17. Ibid., vol. 16, p. 336
18. Ibid., vol. 7, p. 321
19. MacIntyre, Alasdair, *After Virtue: A Study in Moral Theory*, London, 1981, p. 133

PART III: THE MACHINES

1. Kierkegaard, Søren, quoted in Westphal, Merold, *Kierkegaard's Critique of Reason and Society*, Pennsylvania, 1991, p. 117

8. DEMYSTIFYING THE MACHINE

1. http://echo.mpiwg-berlin.mpg.de/ECHOdocuView?url=/mpiwg/online/permanent/archimedes/galil_mecha_070_en_1665
2. Stillman, Drake and Drabkin, I. E. (eds.), *Mechanics in Sixteenth-century Italy. Selections from Tartaglia, Benedetti, Guido Ubaldo, and Galileo*, Madison, 1969, p. 241
3. Smith, Adam, *An Enquiry into the Nature and Causes of the Wealth of Nations*, London, 1776, p. 352
4. https://www.genome.gov/27565109/the-cost-of-sequencing-a-human-genome/
5. https://www.kuka.com/en-gb/industries/metal-industry
6. https://www.ft.com/content/f809870c-26a1-11e7-8691-d5f7e0cd0a16
7. 'Measuring the Internet Economy: A Contribution to the Research Agenda', OECD Digital Economy Papers, no. 226, 2013
8. https://www.theatlantic.com/business/archive/2018/01/amazon-mechanical-turk/551192/
9. http://www.cs.virginia.edu/~robins/Turing_Paper_1936.pdf
10. https://www.cs.washington.edu/events/colloquia/search/details?id=560
11. Turing, Alan, 'Computing machinery and intelligence', *Mind*, vol. LIX, no. 236, October 1950
12. Cobb, Matthew, *Life's Greatest Secret: The Race to Crack the Genetic Code*, London, Kindle edn, p. 23
13. Nadler, Steven, *The Philosopher, the Priest, and the Painter: A Portrait of Descartes*, Princeton, 2013, p. 26
14. http://www.davidhume.org/texts/dnr.html
15. https://archive.org/stream/TheMysteriousUniverseSirJamesJeans/The%20mysterious%20universe%20-%20Sir-James%20Jeans_djvu.txt
16. Zuse, K., *Calculating Space*, Massachusetts Institute of Technology (Project MAC), Cambridge, MA, 1970
17. http://cqi.inf.usi.ch/qic/wheeler.pdf
18. Chaitin, Gregory, *Proving Darwin: Making Biology Mathematical*, New York, 2012, p. 17
19. See for example, Cobb, *Life's Greatest Secret*
20. Weinberg, Steven, 'Is the universe a computer?', *New York Review of Books*, 24 October 2002
21. Gleick, James, *The Information: A History, a Theory, a Flood*, London, 2011, p. 10
22. Squires, Euan J., *The Mystery of the Quantum World*, Boca Raton, 1994

23. Forman, Paul, 'Weimar culture, causality, and quantum theory: Adaptation by German physicists and mathematicians to a hostile environment', *Historical Studies in the Physical Sciences*, vol. 3, 1971, pp. 1–115

24. Wiener, Norbert, *Cybernetics: Or Control and Communication in the Animal and the Machine*, Cambridge, 1948, Kindle edn, p. 132

25. http://www.pitt.edu/~jdnorton/lectures/Rotman_Summer_School_2013/thermo_computing_docs/Landauer_1961.pdf

26. Landauer, Rolf, 'The physical nature of information', *Physics Letters A 217*, 1996, pp. 188–93

27. http://spectrum.ieee.org/computing/hardware/landauer-limit-demonstrated

28. https://data-economy.com/data-centres-world-will-consume-1-5-earths-power-2025/

29. Floridi, Luciano, *The Fourth Revolution: How the Infosphere is Reshaping Human Reality*, Oxford, 2014, p. 96

9. WHY DO WE NEED A THEORY OF HUMANS?

1. Aristotle, *The Politics*, Oxford, Kindle edn, p. 10

2. http://www.perseus.tufts.edu/hopper/text?doc=Perseus%3Atext%3A1999.01.0058%3Abook%3D1%3Asection%3D1253b

3. http://www.rivm.nl/bibliotheek/digitaaldepot/20040108nature.pdf

4. https://www.theguardian.com/environment/2016/aug/29/declare-anthropocene-epoch-experts-urge-geological-congress-human-impact-earth

5. Libet, Benjamin et al, 'Time of conscious intention to act in relation to onset of cerebral activities (readiness-potential): The unconscious initiation of a freely voluntary act', *Brain*, 106, 1983, pp. 623–42

6. Harari, Yuval Noah, *Homo Deus: A Brief History of Tomorrow*, London, 2016, Kindle edn, p. 329

7. Lavazza, Andrea, 'Free will and neuroscience: From explaining freedom away to new ways of operationalizing and measuring it', *Front Hum Neurosci*, 10: 262, 2016

8. Schurger, Aaron et al, 'Neural antecedents of spontaneous voluntary movement: A new perspective', *Trends in Cognitive Science*, vol. 20, issue 2, February 2016, pp. 77–9

9. http://www.psychology.emory.edu/cognition/rochat/Rochat5levels.pdf

10. Piaget, Jean and Inhelder, Bärbel, *The Growth of Logical Thinking from Childhood to Adolescence: An Essay on the Construction of Formal Operational Structures*, New York, 1958

11. Klein, Richard G., 'Archeology and the evolution of human behavior', *Evolutionary Anthropology*, vol. 9, issue 1, 2000
12. https://www.marxists.org/archive/marx/works/1844/manuscripts/labour.htm
13. https://www.marxists.org/archive/marx/works/1844/manuscripts/comm.htm
14. Dunayevskaya, Raya, *Marxism and Freedom: From 1776 to Today*, London, 1971

10. THE THINKING MACHINE

1. https://web.archive.org/web/20140623034804/http://classicgaming.gamespy.com/View.php?view=Articles.Detail&id=395
2. https://arxiv.org/pdf/1312.5602v1.pdf
3. http://www.andreykurenkov.com/writing/a-brief-history-of-neural-nets-and-deep-learning-part-4/
4. http://blog.citizennet.com/blog/2012/11/10/random-forests-ensembles-and-performance-metrics
5. https://www.wired.com/2016/03/two-moves-alphago-lee-sedol-redefined-future/
6. https://www.newscientist.com/article/2080927-how-victory-for-googles-go-ai-is-stoking-fear-in-south-korea/
7. Dick, Philip K., *Do Androids Dream of Electric Sheep?*, Garden City, 1968, p. 29
8. https://deepmind.com/blog/alphago-zero-learning-scratch/
9. https://standards.ieee.org/develop/project/7000.html
10. http://www.fao.org/faostat/en/#data/QC
11. Nietzsche, Friedrich, *The Will to Power*, New York, 1968, p. 962
12. Marx, Karl, 'The Power of Money', *Karl Marx and Friedrich Engels Collected Works*, vol. III, New York, 1975, p. 326
13. https://www.youtube.com/watch?v=fJ2T5FsUI6c
14. https://www.theguardian.com/technology/2017/jul/25/elon-musk-mark-zuckerberg-artificial-intelligence-facebook-tesla
15. https://chinacopyrightandmedia.wordpress.com/2017/07/20/a-next-generation-artificial-intelligence-development-plan/
16. https://www.wired.com/story/for-superpowers-artificial-intelligence-fuels-new-global-arms-race/
17. https://www.ft.com/content/856753d6-8d31-11e7-a352-e46f43c5825d
18. https://www.nytimes.com/2017/09/01/opinion/artificial-intelligence-regulations-rules.html

19. https://deepmind.com/applied/deepmind-ethics-society/research/AI-morality-values/
20. Omohundro, Steve, 'Autonomous technology and the greater human good', *Journal of Experimental & Theoretical Artificial Intelligence*, 26:3, 2014, pp. 303–15
21. Yudkowsky, Eliezer, 'Artificial Intelligence as a positive and negative factor in global risk', in Bostrom, Nick and Circovic, Milan (eds.), *Global Catastrophic Risk*, Oxford, 2008, p. 308
22. http://articles.latimes.com/2005/sep/04/nation/na-levee4
23. https://www.thenation.com/article/undone-neoliberalism/
24. https://www.theguardian.com/technology/2018/mar/28/uber-arizona-secret-self-driving-program-governor-doug-ducey

11. THE ANTI-HUMANIST OFFENSIVE

1. Stapledon, Olaf, *Last and First Men*, London, 2011, Kindle edn, p. 180
2. Wiener, Norbert, *The Human Use of Human Beings*, Boston, 1954, p. 46
3. Huxley, Julian, *New Bottles for New Wine*, London, 1957, pp. 13–17
4. https://nickbostrom.com/papers/history.pdf
5. https://www.themaven.net/transhumanistwager/transhumanism/transhumanism-is-under-siege-from-socialism-UzA2xHZiFUaGOiUFpcon5g/
6. Annas, G. et al, 'Protecting the endangered human: Toward an international treaty prohibiting cloning and inheritable alterations', *American Journal of Law and Medicine*, 28 (2&3), 2002, pp. 151–78
7. Fukuyama, Francis, *Our Posthuman Future: Consequences of the Bio-technology Revolution*, New York, 2002, Kindle Location 2316–17
8. Foucault, Michel, *The Order of Things: An Archaeology of the Human Sciences*, London, 2005, p. 422
9. https://www.marxists.org/reference/archive/althusser/1964/marxism-humanism.htm
10. https://www.marxists.org/reference/archive/althusser/1969/lenin-before-hegel.htm
11. Clarke, Simon et al, *One-Dimensional Marxism: Althusser and the Politics of Culture*, London, 1980, pp. 5–102
12. Baudrillard, Jean, *The Ecstasy of Communication*, Los Angeles, 2012
13. Braidotti, Rosi, *The Posthuman*, Cambridge, 2013, p. 5
14. All quotes in this section:https://web.archive.org/web/20120214194015/http://www.stanford.edu/dept/HPS/Haraway/CyborgManifesto.html

15. http://ieeexplore.ieee.org/document/4065609/
16. Maturana, Humberto R., 'Communication and Representation Functions', Biology. Computer Lab Report 267, University of Illinois, Urbana, 1975
17. Harman, Graham, 'On vicarious causation', *Collapse*, 2, pp. 187–221
18. Coole, Diana and Frost, Samantha (eds.), *New Materialisms: Ontology, Agency, and Politics*, Durham and London, 2010, p. 7
19. Harman, Graham, *Object Oriented Ontology: A New Theory of Everything*, 2018, Kindle edn, p. 144
20. http://www.drps.ed.ac.uk/14-15/dpt/cxartx11039.htm
21. https://edinburghuniversitypress.com/series-new-materialisms.html
22. Coole and Frost, *New Materialisms,* p. 7
23. Latour, Bruno and Woolgar, Steve, *Laboratory Life: The Social Construction of Scientific Facts*, Beverly Hills, 1979. See also, Latour, Bruno and Woolgar, Steve, *Laboratory Life: The Construction of Scientific Facts*, 2nd edn with postscript, Princeton, 1986, p. 273
24. See for example, Flax; Jane, 'Postmodernism and gender relations in feminist theory', *Signs*, vol. 12, no. 4: *Within and Without: Women, Gender, and Theory*, Summer 1987, pp. 621–43
25. Harding, Sandra, *The Science Question in Feminism*, Cornell, 1986
26. Latour, *Laboratory Life: The Construction of Scientific Facts*, p. 280
27. Latour, Bruno, 'Do scientific objects have a history? Pasteur and Whitehead in a bath of lactic acid', *Common Knowledge*, vol. 5, no. 1, Spring 1996
28. Koertge, Noretta, 'Scrutinising science studies', in Koertge, Noretta, *A House Built on Sand: Exposing Postmodernist Myths about Science*, Oxford, 1998
29. Latour, Bruno, 'Why has critique run out of steam? From matters of fact to matters of concern', *Critical Inquiry*, vol. 30, no. 2, Winter 2004, pp. 225–48
30 Hayles, Katherine N., *How We Became Posthuman: Virtual Bodies in Cybernetics, Literature, and Informatics*, Chicago, Kindle edn, p. 2
31. Lukacs, Georg, *The Destruction of Reason*, Delhi, 2016, p. 854
32. Fromm, Erich, *The Anatomy of Human Destructiveness*, New York, 1973, p. 389
33. MacCormack, Patricia, *Posthuman Ethics: Embodiment and Cultural Theory*, London, 2012, p. 144
34. Dossett, Renata D., 'The historical influence of Classical Islam on Western humanistic education', *International Journal of Social Science and Humanity*, vol. 4, no. 2, March 2014

35. http://www.perseus.tufts.edu/hopper/text?doc=Perseus%3Atext%3A1
    999.02.0115%3Aact%3D1#note2
36. Fanon, Frantz, *Black Skin, White Masks*, New York, 1967, Kindle
    Location 292–3

12. THE SNOWFLAKE INSURRECTION

1. Woolf, Virginia, *Mr Bennett and Mrs Brown*, London, 1924, p. 5
2. Ibid.
3. Woolf, Virginia, *Orlando: A Biography*, London, 1928, p. 213
4. Wellman, Barry et al, 'The social affordances of the Internet for networked individualism', *Journal of Computer-Mediated Communication*, vol. 8, issue 3, 1 April 2003
5. Baumeister, R. F., 'The self', in Gilbert, D. T. et al (eds.), *The Hand-
   book of Social Psychology*, vol. 1, Boston, 1998, pp. 680–740
6. Wertheim, Margaret, *The Pearly Gates of Cyberspace: A History of
   Space from Dante to the Internet*, London, 1999
7. Ibid.
8. McConnell, Allen et al, 'The self as a collection of multiple self-aspects:
   Structure, development, operation and implications', *Social Cognition*,
   vol. 30, no. 4, 2012, pp. 380–95
9. McKenna, K. Y. A., Green, A. S. and Gleason, M. J., 'Relationship formation on the Internet: What's the big attraction?', *Journal of Social
   Issues*, 58, 2002, pp. 9–32
10. http://bosworth.ff.cuni.cz/finder/3/false?page=1
11. https://www.eff.org/deeplinks/2017/07/global-condemnation-turkeys-
    detention-innocent-digital-security-trainers
12. https://medium.com/i-data/israel-gaza-war-data-a54969aeb23e
13. http://mondoweiss.net/2014/07/terrifying-tweets-israeli/
14. https://www.theguardian.com/uk-news/2018/mar/20/cambridge-
    analytica-execs-boast-of-role-in-getting-trump-elected
15. https://www.theguardian.com/politics/2017/mar/04/nigel-oakes-
    cambridge-analytica-what-role-brexit-trump
16. https://www.bloomberg.com/news/articles/2016-10-27/inside-the-trump-
    bunker-with-12-days-to-go
17. https://www.wired.com/story/russian-facebook-ads-targeted-us-voters-
    before-2016-election/
18. https://www.thescriptsource.net/Scripts/FightClub.pdf

PART IV: MARX

1. Dunayevskaya, Raya, *Marxism and Freedom*, London, 1971, p. 22

13. BREAKING THE GLASS

1. http://edition.cnn.com/2017/08/11/us/charlottesville-white-nationalists-rally-why/index.html
2. https://sites.google.com/site/breivikmanifesto/2083/introduction/05
3. Jamin, Jerome, 'Cultural Marxism: A survey', *Religion Compass*, vol. 12, issue 1–2, January–February 2018
4. http://foreignpolicy.com/2017/08/10/heres-the-memo-that-blew-up-the-nsc/
5. https://www.marxists.org/archive/marx/works/1844/manuscripts/labour.htm
6. http://www.spiegel.de/international/zeitgeist/is-the-lion-man-a-woman-solving-the-mystery-of-a-35-000-year-old-statue-a-802415.html
7. http://www.loewenmensch.de/index.html
8. https://blog.britishmuseum.org/the-lion-man-an-ice-age-masterpiece/
9. Uomini, Natalie Thaïs and Meyer, Georg Friedrich, 'Shared brain lateralization patterns in language and Acheulean stone tool production: A functional transcranial Doppler ultrasound study', *PLOS ONE*, 8(8), 2013
10. Tomasello, Michael, *A Natural History of Human Thinking*, Harvard, 2014, Kindle edn
11. https://www.marxists.org/reference/archive/feuerbach/works/essence/ec15.htm
12. Tomasello, *A Natural History of Human Thinking*, p. 154
13. MacIntyre Alasdair, 'Breaking the chains of reason', in MacIntyre, Alasdair, *After Virtue: A Study in Moral Theory*, London, 1981, p. 143
14. Gilman, Antonio, 'Explaining the Upper Paleolithic revolution', in Spriggs, Matthew (ed.), *Marxist Perspectives on Anthropology*, Cambridge, 1984
15. Boehm, Christopher, *Hierarchy in the Forest: The Evolution of Egalitarian Behavior*, Harvard, 1999
16. http://artdaily.com/news/22531/The-Guennol-Lioness-Sells-For-57-2-Million#.W8OSfS-ZNBw
17. Porada, Edith, 'A leonine figure of the Protoliterate Period of Mesopotamia', *Journal of the American Oriental Society*, vol. 70, no. 4, October–December 1950, pp. 223–6

18. https://www.marxists.org/archive/marx/works/1845/holy-family/cho6_2.htm
19. https://www.marxists.org/archive/marx/works/1844/manuscripts/comm.htm
20. Harari, Yuval Noah, *Homo Deus: A Brief History of Tomorrow*, London, 2016, p. 457

14. WHAT'S LEFT OF MARXISM?

1. https://www.marxists.org/archive/marx/works/1845/theses/theses.htm
2. https://www.marxists.org/subject/marxmyths/hal-draper/article2.htm
3. https://www.marxists.org/archive/marx/works/1845/holy-family/cho4.htm
4. Marx, Karl, *Capital*, vol. 1, pp. 171–2
5. https://endofcapitalism.com/2013/05/29/a-feminist-critique-of-marx-by-silvia-federici/
6. Mason, Paul, *Live Working or Die Fighting*, London, 2007
7. https://www.marxists.org/archive/marx/works/1877/anti-duhring/cho7.htm
8. https://www.marxists.org/archive/trotsky/1938/morals/morals.htm
9. Gorz, André, *Critique of Economic Reason*, London, 1989, p. 8
10. https://operaismoinenglish.files.wordpress.com/2013/06/factory-and-society.pdf
11. See Mason, Paul, *Postcapitalism: A Guide to Our Future*, London, 2015
12. http://thenewobjectivity.com/pdf/marx.pdf
13. See for example, Pitts, Frederick Harry, *Critiquing Capitalism Today: New Ways to Read Marx*, Springer International Publishing, Kindle Location 3953
14. https://www.marxists.org/archive/marx/works/1859/critique-pol-economy/preface.htm
15. Kepler, Johannes, *A New Year's Gift, or On the Six Cornered Snowflake*, Frankfurt am Main, 1611
16. Gare, Arran, 'Aleksandr Bogdanov and Systems Theory', *Democracy and Nature*, vol. 6, p. 3, 2000
17. https://www.marxists.org/archive/marx/works/1844/manuscripts/comm.htm
18. https://www.marxists.org/archive/marx/works/1894-c3/ch46.htm
19. Marx, *Capital*
20. Thompson, E. P., *The Poverty of Theory and Other Essays*, London, 1988, p. 188

PART V: SOME REFLEXES

1. MacIntyre, Alasdair, 'The algebra of the revolution' (1958), in Blackledge, Paul and Davidson, Neil (eds.), *Alasdair MacIntyre's Engagement with Marxism*, Chicago, 2009, p. 44

15. UN-CANCEL THE FUTURE

1. Fisher, Mark, *Ghosts of My Life*, Arlesford, 2013, p. 49
2. https://www.theguardian.com/commentisfree/2016/apr/24/the-new-feudalism-silicon-valley-overlords-advertising-necessary-evil

16. REACT TO THE DANGER

1. Orwell, George, *The Orwell Diaries*, London, 2018, Kindle Location 5739–40
2. Ibid., Kindle Location 5585–7
3. Kelly, John, *Never Surrender: Winston Churchill and Britain's Decision to Fight Nazi Germany in the Fateful Summer of 1940*, London, 2015, p. 251
4. http://www.rightwingwatch.org/post/roy-moore-boasts-of-endorsements-from-neo-confederate-secessionist-activist-who-says-its-ok-to-murder-abortion-providers/
5. http://www.liberation.fr/france/2014/02/06/le-6-fevrier-1934-un-mythe-fondateur-de-l-extreme-droite_978118

17. REFUSE MACHINE CONTROL

1. Jaspers, Karl, 'The Axial Age of human history: A base for the unity of mankind', *Commentary*, 1 November 1948
2. Graeber, David, *Debt: The First 5,000 Years*, Brooklyn, 2011, Kindle Location 5224–7
3. https://www.youtube.com/channel/UCOD61z-vlytlhJDKje5zNPw
4. Bridle, James, 'Something is wrong on the Internet', *Medium*, 6 November 2017
5. https://tcta.org/node/13251-issues_with_test_based_value_added_models_of_teacher_assessment
6. https://www.ted.com/playlists/171/the_most_popular_talks_of_all

7. https://moneysavingexpert.com/students/repay-post-2012-student-loan/

8. 'Policymakers around the world are embracing behavioural science', *Economist*, 18 May 2017

9. http://38r8om2xjhhl25mw24492dir.wpengine.netdna-cdn.com/wp-content/uploads/2017/03/Encouraging_people_into_university.pdf

18. REJECT THE THOUGHTS OF XI JIN PING

1. https://www.wsj.com/articles/america-first-doesnt-mean-america-alone-1496187426

2. http://rukor.org/seven-deadly-sins-in-todays-china/

3. Qian Gang, 'Reading Chinese Politics in 2014', China Media Project, 30 December 2014

4. Suisheng, Zhao, 'The ideological campaign in Xi's China rebuilding regime legitimacy', *Asian Survey*, 56(6), November 2016, pp. 1168–93

5. https://uk.reuters.com/article/uk-china-politics/chinas-xi-says-study-capitalism-but-marxism-remains-top-idUKKCN1C5034

6. https://publishing.cdlib.org/ucpressebooks/view?docId=ft0489n683&chunk.id=d0e177&toc.depth=1&toc.id=d0e177&brand=ucpress

7. http://cpcchina.chinadaily.com.cn/2010-10/20/content_13918219.htm

8. http://www.wired.co.uk/article/chinese-government-social-credit-score-privacy-invasion

19. NEVER GIVE IN

1. Michel, Louise, *Red Virgin: Memoirs of Louise Michel*, Alabama, 2003, p. 112

2. Saussol, Alain, *L'héritage: Essai sur le problème foncier mélanésien en Nouvelle-Calédonie*, Paris, 1979

3. Boime, Albert, *Art in an Age of Civil Struggle, 1848–1871*, Chicago, 2008, Kindle Location 10361

4. https://www.marxists.org/archive/marx/works/1871/civil-war-france/ch05.htm

5. See for example, Tombs, Robert, 'Warriors and killers: Women and violence during the Paris Commune 1871', in Aldrich, R. and Lyons, M. (eds.), *The sphinx in the Tuileries and Other Essays in Modern French History*, Sydney, 1999, pp. 169–82

20. LIVE THE ANTIFASCIST LIFE

1. Foucault, Michel, 'Introduction', in Deleuze, Gilles, *Anti-Oedipus*, Minnesota, 1983, p. xiii

2. Foucault, Michel, 'The Ethics of Care of the Self as Practice of Freedom', in Bernauer, James and Rasmussen, David (eds.), *The Final Foucault*, Cambridge, 1987

3. Orwell, George, *Homage to Catalonia*, London, 1938, p. 1

4. http://orwell.ru/library/essays/Spanish_War/english/esw_1

5. http://eljanoandaluz.blogspot.com/2017/10/voluntarios-internacionales-en-las.html

6. https://www.marxists.org/history/etol/revhist/backiss/vol5/no4/casci ola4.html

7. https://www.orwellfoundation.com/the-orwell-foundation/orwell/poetry/the-italian-soldier-shook-my-hand/

8. http://www.slate.com/articles/news_and_politics/fascism/2017/01/how_italian_fascists_succeeded_in_taking_over_italy.html

9. Thompson, E.P., 'Eighteenth-century English society: Class struggle without class?', *Social History*, vol. 3, no. 2, May 1978, pp. 133–65

10. https://www.redpepper.org.uk/we-are-the-crisis-of-capital/

11. MacIntyre, Alasdair, 'The algebra of the revolution' (1958), in Blackledge, Paul and Davidson, Neil (eds.), *Alasdair MacIntyre's Engagement with Marxism*, Chicago, 2009

Acknowledgements

In writing this book I've been helped by the following people. Joana Ramiro deserves star billing as my researcher. Tom Penn, my editor, invested huge amounts of time and belief; my agent Matthew Hamilton and the team at Aitken Alexander kept me going. The De-growth Institute at Jena University in October 2017 held a seminar focused on challenges to my postcapitalism thesis. Dr Emma Dowling, a researcher at Jena, guided me towards important texts. A postgraduate seminar organized by Professor Adam Morton at the Political Economy department of the University of Sydney discussed the 'General Theory of Trump' in February 2017. Jamie Dobson, CEO of Container Solutions, helped me understand the engineering challenges of AI, while staff at DeepMind helped me to understand how artificial neural networks work. Elena Massa facilitated and translated during the Catalan independence uprising while Eoin Ó Broin, a Sinn Fein TD, helped me understand what was going on. The argument of Chapter 5 was influenced by my participation in the theatre piece *Why It's Kicking Off Everywhere*, jointly devised with Young Vic Theatre artistic director David Lan, producer Ben Cooper and actors Khaled Abdallah, Sirine Saba and Lara Sawalha in April 2017. Hall Greenland and Sue Burrows facilitated my research trip to Sydney. Calum Walton and Jane Bruton gave me realtime feedback on the first draft. Dr Karin Speedy translated and facilitated on my trip to New Caledonia. Theatre-maker Samantha Jayne Williams worked with me on an interactive game I've used to better understand the failure of globalization. The research process has taken me to numerous European countries, with the help of the Biennale Warszawa, the Volksbuhne in Berlin, the Rosa Luxemburg Foundation, the

Friedrich Ebert Stiftung, Das Progressive Zentrum, the Ferdinand Lassalle Institute, Centro Cultural Belem, the Museo Reina Sofia in Madrid and the Karl Renner Institut. The South Korean magazine *SISA-IN* facilitated my travel to Seoul to discuss the ethical challenge of AI with practitioners there. Evgeny Morozov, Francesca Bria, Srećko Horvat, Nadia Idle, Paul Greengrass, Paul Hilder, Matthew Cobb, Ewa Jasiewicz, Aaron Bastani, Theopi Skarlatos and Terry Eagleton had conversations with me that produced lightbulb moments. All the mistakes are mine.

Index

tech monopolies, 18, 40, 73, 143, 202, 251–2, 269
see also financial crisis, global (from 2007/8); globalization
global politics
absence of progress in neoliberal era, 248
bipolar response to disorder, 8
China's reaction to Trump, 277–8, 279–80
and collapse of neoliberalism, xii–xiii, 7–8, 72–5, 82
collapse of Soviet Bloc (1989–91), 46–7, 53
elite responses to networked protest, 198–203
era of de-globalization, xiii, 7–8, 15–16, 100
era of right-wing nationalism, xii–xiii, 5–6, 7–8, 11–12, 15–34, 80–93, 97–100, 188, 198–205, 256–62
geopolitical strategy of Putin, 33, 62, 89–93
global protest movement (from mid-2009), 74–9, 93, 192–3, 198–203, 210
lack of neoliberal geopolitical system, 58, 59
major powers' AI strategies, 160–61
Obama's 'pivot' towards Asia, 91, 279
Obama's soft isolationism, 32
post-9/11 US foreign policy, 61–2
Putin's attacks on multilateral institutions, 90–91, 92–3
Putin's invasion of Georgia (2008), 89–90
Putin–Trump links, 7, 16, 31–3, 92

right-wing inter-operation and support, 7–8, 16, 31–4, 92–3, 189, 201, 202
rule-free environments, 92
Trump's attacks on multilateral institutions, 11, 15, 16, 277–8
Trump's deliberate disorder strategy, 11, 277–8
US as unrivalled superpower, 59, 61–2
globalization
belief in permanence of, 58, 61–2, 91
collapse of, xiii, 5, 8, 15–16, 23–4, 72–5, 100
elites and middle classes of emerging markets, 73–4, 252
as neoliberalism's second phase, 41, 46–7, 53
and Tea Party movement, 18
technocratic model of, 31
Trump as wrecking ball against, 11, 15, 16, 277–8
US imposition/shaping of, 23, 91
Go (game), 147–8, 149–50, 157
Google, 31, 73, 116, 147, 234, 269
collusion with Trump's election campaign, 33, 201, 202
Gorz, André, 233
Graeber, David, 251, 265–6
Greece
2015 referendum, 15
classical age, 265, 294
neoliberal destruction of, 70, 71, 288, 289
occupation and mass protests (2011–12), 76, 78, 288
Syriza in, 262, 288
Greenberg, Stan, 18
Greenspan, Alan, 55, 60, 61, 118
Greenwald, Glenn, 200

labour
  class and surplus product,
    217–18, 219–20, 226
  domestic labour, 227–8
  female participation in
    workforce, 85, 86
  Graeber's 'bullshit jobs', 251
  and history of automation, xi–
    xii, 235–6
  increased global workforce,
    46–7, 72
  industrial capitalism of
    nineteenth-century, 186–7,
    219–20, 224–5, 226
  'labour theory of value', 115–16,
    118, 226, 235
  Marx's theories, 118, 140–41,
    213–15, 217–18, 219–20,
    222, 226, 227–8, 235–6
  slide in 'wage share' of developed
    economies, 71
labour movements, 37–9, 41, 42,
    53, 65, 177, 274
  in the 1920s, 298–9
  early twentieth century revolts,
    109, 191–2
  male trade unionism, 294
  neoliberal destruction of,
    38–9, 40, 41, 42, 46, 53,
    177, 187, 233
  virtues and morality, 205,
    231, 232–3
Labour Party, British, 203, 261,
    262, 263, 272, 298
Landauer, Rolf, 128
language, 138, 214–15, 216, 220–21
Lao-tse, 265
Lapavitsas, Costas, 54
Laplace, Pierre-Simon, 122
Latour, Bruno, 183, 184–6, 188
Laurat, Lucien, 107

Lavazza, Andrea, 137
Law and Justice Party, Poland, 262
Lawrence, D. H., *The Daughter-in-
    Law*, 191
Le Pen, Marine, 294
League of the South, 258–9
Lebanon, 90
Lee Sedol, 148, 149–50, 157
left-wing thought
  academic attacks on humanism,
    xiii, 12–13, 171, 174–7,
    188–9, 221, 241
  anti-Bolshevik tradition, 106–8
  'authoritarian personality'
    theory, 98, 99
  British Labour Party, 203, 261,
    262, 263, 272, 298
  Corbyn's radical Labour
    manifesto (2017), 262, 272
  critique of science, 12–13, 171,
    174–5, 183–6, 187, 188
  and defence of rule of law, 260–61
  economics as help against
    far-right, 261–2
  fragmentations and dualism,
    174–5, 177, 178
  inability to project clear
    alternative, 68
  'knowing what kind of world we
    live in', 256–8, 261, 263
  Leninist theory of party and
    revolution, 176
  need for unity against fascism,
    259–60, 297–8
  need to work with science, 238
  networked-activists' move into
    party politics, 263, 298
  New Left, 178–9
  new left parties, 262–3, 288
  no return to state socialism,
    254, 255

monopolies
aligned to state power, 18, 62,
143, 252, 271
and Iraq War, 62
tech companies, 18, 40, 73, 143,
202, 251–2, 253, 255, 269
del Monte, Guidobaldo, 115
Moore, Roy, 258–9
Morozov, Evgeny, 249
Moss, Kate, 247, 249
Mubarak, Gamal, 76
Mubarak, Hosni, 75, 78
Murdoch, Rupert, 89, 193
Murray, Charles, *The Bell Curve*
(1994), 95, 96
music, digital, 117, 128
Musk, Elon, 160
Mussolini, Benito, 107, 187,
296, 297

NAFTA, 11
Nagle, Angela, 83–4
Napoleon, 56–7
nationalism, authoritarian
anti-EU alliances, 27
evolution towards fascism, 98–9,
258–9
'national neoliberalism', 23–4,
30, 71, 72–3
Putin in Russia, 11
rise of in twenty-first century,
xii–xiii, 5–6, 7–8, 11–12,
15–34, 80–93, 97–100, 188,
198–205, 256–62
Trump's economic nationalist
narrative, 17, 24, 26, 33,
71, 98–9
see also right-wing
authoritarianism
NATO, 89, 91
Nemtsov, Boris, 93

neoliberal self, 49
'borrowing is good' lesson,
50–52, 53–4, 63
consumption as self-validating
activity, 49, 52–3, 59, 63
crisis of, 39, 55, 66–7, 68–71, 73,
74–9, 192–3
damage to collective human
psyche, 34, 41–6, 48–50,
52–4, 63, 64, 66–7, 100
and free-market ideology, 8,
12–13, 34, 38–9, 41, 64, 82,
142–3, 187
in new industrial heartlands, 230
performative nature of, 16–17,
67, 71, 100, 165, 228, 234,
299–300
and postmodernism, 12–13,
174–5, 177, 187
the rootless and self-centred
individual, 45–6, 48–50
neoliberalism
first phase of (1979–89), 41–6
second phase of (1989–2001), 41,
46–55, 57–60
third phase of (2001–08), 41, 55,
60–63, 64
fourth phase of (2008–16), 41,
66–7, 68–71, 72–5
and America's black
communities, 94, 95
and behavioural psychology, 165,
270–73
'capitalist communism' of, 23
collapse of, xii–xiii, 7–8, 39, 58,
68–79, 82–9, 93–101, 109,
228–9, 233–4, 256–7, 259
see also financial crisis, global
(from 2007/8)
co-option of ethical
consumption, 273